Muscle Energy Techniques

For Churchill Livingstone:

Commissioning editor: Inta Ozols
Project development editor: Valerie Bain
Project manager: Valerie Burgess
Project controller: Pat Miller/Derek Robertson
Design direction: Judith Wright
Copy editor: Stephanie Pickering
Sales promotion executive: Maria O'Connor

Muscle Energy Techniques

Rules:

Leon Chaitow ND DO
Practitioner and Senior Lecturer, University of Westminster, London, UK

Craig Liebenson DC (Chapter 5: Manual Resistance Techniques and rehabilitation)
Private Practitioner, Los Angeles, USA

Illustrated by
Graeme Chambers BA(Hons)
Medical Artist

Foreword by
Professor Laurie Hartman
British School of Osteopathy, London, UK

Series Editor
Professor Patrick C Pietroni FRCGP MRCP DCH
Director of the Centre for Community Care and Primary Health, University of Westminster, London, UK

CHURCHILL
LIVINGSTONE

EDINBURGH LONDON NEW YORK PHILADELPHIA ST LOUIS SYDNEY TORONTO 1996

CHURCHILL LIVINGSTONE
An imprint of Harcourt Publishers Limited

D is a registered trademark of Harcourt Publishers
Limited

First published 1996
    Reprinted 1997
    Reprinted 1998
    Reprinted 1999
    Reprinted 2000

ISBN 0 443 05297 2

British Library Cataloguing in Publication Data
A catalogue record for this book is available from the British
Library.

Library of Congress Cataloging in Publication Data
A catalogue record for this book is available from the
Library of Congress.

Medical knowledge is constantly changing. As new
information becomes available, changes in treatment,
procedures, equipment and the use of drugs become
necessary. The editors/authors/contributors and the
publishers have, as far as it is possible, taken care to ensure
that the information given in this text is accurate and up to
date. However, readers are strongly advised to confirm that
information, especially with regard to drug usage, complies
with the latest legislation and standards of practice.

Neither the publishers nor the author will be liable for any
loss or damage of any nature occasioned to or suffered by
any person acting or refraining from acting as a result of
reliance on the material contained in this publication.

The
publisher's
policy is to use
**paper manufactured
from sustainable forests**

Printed in China
NPCC/05

Contents

Abbreviations

AC: acromioclavicular
ASIS: anterior superior iliac spine
ATP: adenosine triphosphate
CNS: central nervous system
CRAC: contract–relax, antagonist–contract
EMG: electromyograph
FMS: fibromyalgia syndrome
FPR: facilitated positional release
INIT: integrated neuromuscular inhibition technique
MET: muscle energy technique
MPS: myofascial pain syndrome
MRT: manual resistance technique

NMT: neuromuscular technique
OMT: osteopathic manipulative therapy
PIR: postisometric relaxation
PNF: proprioceptive neuromuscular facilitation
PSIS: posterior superior iliac spine
QL: quadratus lumborum
RI: reciprocal inhibition
SCM: sternocleidomastoid
SCS: strain/counterstrain
SI: sacroiliac
TFL: tensor fascia lata
TMJ: temperomandibular joint
VMO: vastus medialis oblique

Foreword

In general, osteopathic technique evolves steadily over the years. However, every so often, a major leap forward occurs, either because of the ability of one or more individuals who have a wider view, or due to a natural progression as experience highlights a gap in current methods.

The original development of muscle energy technique by Fred Mitchell Snr was one of these major changes in practice: as classified and taught by him it has stood the test of time and developed into an extremely useful approach for many bodywork therapists. The fact that the method has been gradually adopted not only by osteopaths but also by many of the other professions using manual approaches, is a mark of its usefulness.

Recently, interest in the use of indirect techniques has increased as many practitioners move away from confrontational techniques towards what are possibly less invasive procedures. There are many reasons for this change, not least the fear of litigation when forceful techniques are performed badly and severe adverse reactions occur. This is a pity, as well-applied, conventional manipulative techniques should not cause pain and can be extremely effective in bringing about the rapid resolution of many pain syndromes. Also, importantly, indirect techniques, of which muscle energy is just one approach, are far less likely to produce severe or long-lasting adverse reactions in less well-trained hands. Not that this is an excuse for inefficient or poorly applied technique, as the responsibility to the patient is not merely to do no harm but to produce the quickest and best result.

A welcome feature of this book is the detailed way in which it shows the practical application in conjunction with the reasoning, enabling the student and practitioner to use the system immediately in a constructive way. Further integration is brought to the approach by the way methods are combined seamlessly with other technique systems, making treatment more efficient for the operator and more effective for the patient. Extensive referencing pays tribute to the author's wide knowledge and tireless research, and shows the depth to which he has investigated his subject to verify the suppositions made in the descriptions and the proposed purpose of the technique.

As students of the various disciplines of manual treatment are required to attain ever higher standards of education and investigative thinking, more detail about reasoning and background is needed to satisfy the questing mind. The problem of translating this intelligence into a reproducible therapeutic modality that can be learnt, modified and utilised, is considerable. Leon Chaitow is to be congratulated for the exceptionally logical way in which he has combined the thinking with the doing. The reasoning behind the technique is clearly important, but if the diagrams and instructions are less than useful then the whole system falls flat. This is not likely to be the case with this work!

The author also painstakingly points out the advantages, from several viewpoints, of adopting this method. The use of manual methods of treatment is, in fact, growing in acceptance in many countries of the world, but the quantity of published material is increasing very slowly considering the developing interest in the subject within the medical and supplementary professions. This book should help to assist in further promulgating information on the work done by qualified practitioners. Its scope is wide, but it is designed to make the subject as simple as possible bearing in mind the complexity of the subject.

1996   L.H.

Acknowledgements

My sincere thanks go to the many manual medicine experts, osteopaths, chiropractors, physiotherapists and others, whose work I have drawn on for this book. In particular, I wish to pay tribute to the generous and warm collaboration of Dr Craig Liebenson, who has done so much to bring these methods into mainstream chiropractic, as well as to the two giants of manual medicine, Karel Lewit and Vladimir Janda, who have given so many insights and so much impetus to the evolution of osteopathic 'muscle energy techniques'. Within that profession, Drs Korr, Greenman, Stiles, Mitchell and Patterson are deserving of special mention, as are Drs Travell and Simons for their work in myofascial dysfunction. My thanks too go to Aaron Mattes for showing me new ways to stretch!

I dedicate this book to Alkmini and Sasha, who endured my enslavement to the process involved in the writing of the book with extraordinary grace.

1996   L.C.

1

An introduction to muscle energy techniques

MUSCLE ENERGY TECHNIQUES (MET)

A revolution has taken place in manipulative therapy involving a movement away from high velocity/low amplitude thrusts (characteristic of most chiropractic and, until recently, much osteopathic manipulation) towards gentler methods which take far more account of the soft tissue component (DiGiovanna 1991, Greenman 1989, Travell & Simons 1992).

One such approach – which targets the soft tissues primarily, although it also makes a major contribution towards joint mobilisation – has been termed muscle energy technique (MET) in osteopathic medicine. There are a variety of other terms used to describe it, the most general and accurately descriptive of which being 'active muscular relaxation techniques'(Liebenson 1989, 1990). MET is the product of a variety of schools, although its origins may be found in ortho-paedic and physiotherapy techniques as well as in osteopathic work. The current interest in MET methods crosses all such political and thera-peutic barriers.

MET, as presented in this book, owes most of its development to osteopathic clinicians such as Fred Mitchell, Snr. (Mitchell 1967), with more recent refinements deriving from the work of people such as Karel Lewit (Lewit 1986a) and Vladimir Janda (Janda 1989) of the former Czechoslovakia, both of whose work will be referred to many times in this text.

Fred Mitchell Snr.

No single individual deserves credit for MET, but its inception into osteopathic work must be

1

credited to F. L. Mitchell Snr., in 1958. Since then his son (Mitchell et al 1979), and many others, have evolved a highly sophisticated system of manipulative methods (F. Mitchell Jr., tutorial on biomechanical procedures, American Academy of Osteopathy, 1976) in which the patient 'uses his/her muscles, on request, from a precisely controlled position in a specific direction, against a distinctly executed counterforce'.

Philip Greenman's view

Professor of Biomechanics, Philip Greenman states (Greenman 1989),

The function of any articulation of the body which can be moved by voluntary muscle action, either directly or indirectly, can be influenced by muscle energy procedures . . . Muscle energy techniques can be used to lengthen a shortened, contractured or spastic muscle; to strengthen a physiologically weakened muscle or group of muscles; to reduce localised edema, to relieve passive congestion, and to mobilise an articulation with restricted mobility.

Sandra Yale's view

Osteopathic physician Sandra Yale (in: DiGiovanna 1991) extols MET's potential in even fragile and severely ill patients:

Muscle energy techniques are particularly effective in patients who have severe pain from acute somatic dysfunction, such as those with a whiplash injury from a car accident, or a patient with severe muscle spasm from a fall. MET methods are also an excellent treatment modality for hospitalised or bedridden patients. They can be used in older patients who may have severely restricted motion from arthritis, or who have brittle osteoporotic bones.

Edward Stiles

Among the key MET clinicians is Edward Stiles, who elaborates on the theme of the wide range of MET application (Stiles 1984a, 1984b). He states:

Basic science data suggests the musculoskeletal system plays an important role in the function of other systems. Research indicates that segmentally related somatic and visceral structures may affect one another directly, via viscerosomatic and somaticovisceral reflex pathways. Somatic dysfunction may increase energy demands, and it can affect a wide variety of bodily processes; vasomotor

control, nerve impulse patterns (in facilitation), axionic flow of neurotrophic proteins, venous and lymphatic circulation and ventilation. The impact of somatic dysfunction on various combinations of these functions may be associated with myriad symptoms and signs. A possibility which could account for some of the observed clinical effects of manipulation.

As to the methods of manipulation he now uses clinically, Stiles states that he employs muscle energy methods on about 80% of his patients, and functional techniques (such as strain/counterstrain) on 15 to 20%. He uses high velocity thrusts on very few cases. The most useful manipulative tool available is, he maintains, muscle energy technique.

Early sources – going beyond proprioceptive neuromuscular facilitation (PNF)

It is fair to credit the osteopathic profession with most of the recent developments of these methods, since many applications and refinements have sprung from that school of medicine. Earlier, however, a technique of a similar, if simpler, type was employed by physiotherapists who, among other terms, described this more limited version as Proprioceptive Neuromuscular Facilitation (PNF). The method tended to stress the importance of rotational components in the function of joints and muscles, and made use of this by a resisted (isometric) exercise, usually involving extremely strong contractions. Initially, the focus of PNF was the strengthening of neurologically weakened muscles, with attention to the release of muscle spasticity following on from this (Kabot 1959, Levine et al 1954). Mitchell's adaptation of this work for use in joint mobilisation and release of muscle shortness, was a natural evolution which has continued in physiotherapy, manual medicine, osteopathy and, increasingly, in massage therapy and chiropractic settings.

Postisometric relaxation (PIR) and reciprocal inhibition (RI) – two forms of MET

A term much used in more recent developments of these methods is postisometric relaxation

(PIR), especially in relation to the work of Karel Lewit. The term refers to the effect of the subsequent relaxation experienced by a muscle, or group of muscles, after brief periods during which an isometric contraction has been performed.

The terms proprioceptive neuromuscular facilitation (PNF) and postisometric relaxation (PIR) (the latent hypotonic state of a muscle following isometric activity), represent variations on the same theme. A further variation involves the physiological response of the antagonists of a muscle which has been isometrically contracted – reciprocal inhibition (RI).

When a muscle is isometrically contracted, its antagonist will be inhibited, and will relax immediately following this. Thus the antagonist of a shortened muscle, or group of muscles, may be isometrically contracted, in order to achieve a degree of ease and additional movement in the shortened tissues.

Sandra Yale (in: DiGiovanna 1991) acknowledges that, apart from the well understood processes of reciprocal inhibition, the precise reasons for the effectiveness of MET remain unclear – although in achieving PIR the effect of a sustained contraction on the Golgi tendon organs seems pivotal, since their response to such a contraction seems to be to set the tendon and the muscle to a new length by inhibiting it

(Moritan et al 1987). Other terms which have been applied to such methods include 'hold–relax' technique and 'contract–relax' technique, as well as a more recent arrival, which has a distinctly American ring, 'myokinesis'.

Lewit and Simons (1984) agree that, whilst reciprocal inhibition is a factor in some forms of therapy related to postisometric relaxation techniques, it is not a factor in PIR itself, which is a phenomenon resulting from a neurological loop involving the Golgi tendon organs (Figs 1.1 and 1.2).

Where pain of an acute or chronic nature makes controlled contraction of the muscles involved difficult, the therapeutic use of the antagonists can patently be of value. Thus modern MET incorporates both postisometric relaxation and reciprocal inhibition methods, as well as aspects unique to itself, such as isokinetic techniques, described later (p. 62).

A number of researchers, including Karel Lewit of Prague (Lewit 1991), have reported on the usefulness of aspects of MET in the treatment of trigger points, and this is seen by many to be an excellent method of treating these myofascial states, and of achieving the restoration of a situation where the muscle in which the trigger lies is once more capable of achieving its full resting length, with no evidence of shortening.

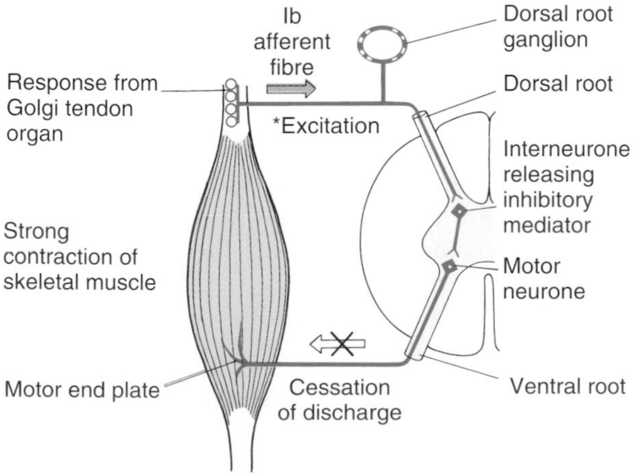

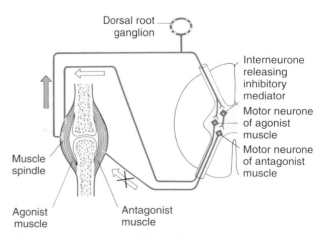

Figure 1.1   Schematic representation of the neurological effects of the loading of the Golgi tendon organs of a skeletal muscle by means of an isometric contraction, which produces a postisometric relaxation effect in that muscle.

Figure 1.2   Schematic representation of the reciprocal effect of an isometric contraction of a skeletal muscle, resulting in an inhibitory influence on its antagonist.

KEY POINTS ABOUT MET

MET methods all employ variations on a basic theme. This primarily involves the use of the patient's own muscular efforts in one of a number of ways, usually in association with the efforts of the therapist:

1. The operator's force may exactly match the effort of the patient (so producing an isometric contraction) allowing no movement to occur – and producing as a result a physiological neurological response (via the Golgi tendon organs) involving a combination of:
— reciprocal inhibition of the antagonist(s) of the muscle(s) being contracted, as well as
— postisometric relaxation of the muscle(s) which are being contracted.

2. The operator's force may overcome the effort of the patient, thus moving the area or joint in the direction opposite to that in which the patient is attempting to move it (this is an isotonic eccentric contraction, also known as an isolytic contraction).

3. The operator may partially match the effort of the patient, thus allowing, although slightly retarding, the patient's effort (and so producing an isotonic concentric contraction).

Other variables may be also introduced, for example involving:

- Whether the contraction should commence with the muscle or joint held at the resistance barrier or short of it – a factor decided largely on the basis of the degree of chronicity or acuteness of the tissues involved
- How much effort the patient uses – say, 20% of strength, or more, or less
- The length of time the effort is held – 7 to 10 seconds, or more, or less
- Whether, instead of a single maintained contraction, to use a series of rapid, low-amplitude contractions (Ruddy's 'rhythmic resisted duction' method, also known as 'pulsed muscle energy technique')
- The number of times the isometric contraction (or its variant) is repeated
- The direction in which the effort is made – towards the resistance barrier or away from it

(thus involving attention to the antagonists to the muscles; or the actual muscles (agonists) which require 'release' and subsequent stretching) – these variations are also known as 'direct' and 'indirect' approaches (p. 11)
- Whether to incorporate a held breath and/or specific eye movements to enhance the effects of the contraction
- What sort of resistance is offered (for example, by the operator, by gravity, by the patient himself, or by an immovable object)
- Whether the patient's effort is matched, overcome or not quite matched – a decision based on the precise needs of the tissues – to achieve relaxation, reduction in fibrosis or tonifying/re-education
- Whether to take the muscle or joint to its new barrier following the contraction, or whether or not to stretch the area/muscle(s) beyond the barrier – this decision is based on the nature of the problem being addressed (does it involve shortening? fibrosis?) and its degree of chronicity
- Whether any subsequent (to a contraction) stretch is totally passive, or whether the patient should participate in the movement, the latter being thought by some to be desirable in order to reduce danger of stretch reflex activation (Mattes 1990)
- Whether to utilise MET alone, or in a sequence with other modalities such as the positional release methods of 'strain/counterstrain', or the ischaemic compression/inhibitory pressure techniques of Neuromuscular Technique (NMT) – such decisions will depend upon the type of problem being addressed, with myofascial trigger point treatment frequently benefiting from such combinations (Chaitow 1993).

Greenman summarises the requirements for the successful use of MET in osteopathic situations as 'control, balance and localisation'. His suggested basic elements of MET include the following:

- A patient/active muscle contraction
- This commences from a controlled position
- The contraction is in a specific direction

(towards or away from a restriction barrier)

- The operator applies distinct counterforce (to meet, not meet, or to overcome the patient's force)
- The degree of effort is controlled (sufficient to obtain an effect but not great enough to induce trauma or difficulty in controlling the effort).

What is done subsequent to the contraction may involve any of a number of variables, as will be explained (pp 7–8).

The essence of MET then is that it uses the energy of the patient, and that it may be employed in one or other of the manners described above with any combination of variables depending upon the particular needs of the case.

Using agonist or antagonist?

As mentioned, a critical consideration in MET, apart from degree of effort, duration and frequency of use, involves the direction in which the effort is made. This may be varied, so that the operator's force is directed towards overcoming the restrictive barrier (created by a shortened muscle, restricted joint, etc.); or indeed opposite forces may be used, in which the operator's counter-effort is directed away from the barrier.

There is general consensus amongst the various osteopathic experts already quoted that the use of postisometric relaxation is more useful than reciprocal inhibition in normalising hypertonic musculature. This, however, is not generally held to be the case by experts such as Lewit and Janda, who see specific roles for these variations (pp 7–8).

Osteopathic clinicians such as Stiles and Greenman believe that the muscle which requires stretching – the agonist – should be the main source of 'energy' for the isometric contraction, and suggest that this achieves a more significant degree of relaxation, and so a more useful ability to subsequently stretch the muscle, than would be the case were the relaxation effect being achieved via use of the antagonist i.e. using reciprocal inhibition.

Following on from an isometric contraction – whether agonist or antagonist is being used – there is a refractory, or latency, period of approximately 15 seconds during which there can be an easier (due to reduced tone) movement towards the new position (new resistance barrier) of a joint or muscle.

Variations on the MET theme

Liebenson (1989, 1990) describes three basic variations which are used by Lewit and Janda as well as by himself in a chiropractic rehabilitation setting.

Lewit's modification of MET, which he calls postisometric relaxation, is directed towards relaxation of hypertonic muscle, especially if this relates to reflex contraction or the involvement of myofascial trigger points. Liebenson (Lewit 1986b) notes that 'this is also a suitable method for joint mobilisation when a thrust is not desirable'.

Lewit's postisometric relaxation method

1. The hypertonic muscle is taken, without force or 'bounce', to a length just short of pain, or to the point where resistance to movement is first noted (Fig. 1.3).

2. The patient gently contracts the affected hypertonic muscle away from the barrier – i.e. the agonist is used – for between 5 and 10 seconds, while the effort is resisted with an exactly equal counterforce. Lewit usually has the patient inhale during this effort.

3. This resistance involves the operator holding the contracting muscle in a direction which would stretch it, were resistance not being offered.

4. The degree of effort, in Lewit's method, is minimal. The patient may be instructed to think in terms of using only 10 or 20% of his available strength, so that the manoeuvre is never allowed to develop into a contest of strength between the operator and the patient.

5. After the effort, the patient is asked to exhale and to let go completely, and only when this is achieved is the muscle taken to a new

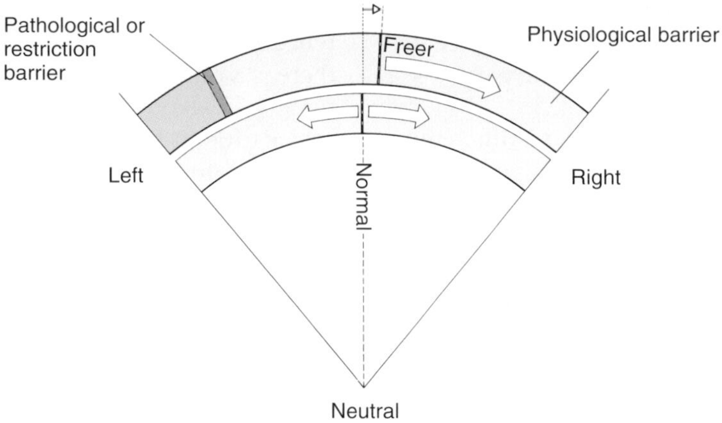

Figure 1.3   A schematic representation of the directions in which a muscle or joint can move – towards a restriction barrier (at which point MET could be usefully applied) or towards a position of relative ease.

barrier with all slack removed but no stretch – to the extent that the relaxation of the hypertonic muscles will now allow.

6. Starting from this new barrier, the procedure is repeated 2 or 3 times more.

7. In order to facilitate the process, especially where trunk and spinal muscles are involved, Lewit usually asks the patient to assist by looking with his eyes in the direction of the contraction during the contracting phase, and in the direction of stretch during the stretching phase of the procedure.

The key elements in this approach, as in most MET, involve precise positioning, as well as taking out slack and using the barrier as the starting and ending points of each contraction.

What is happening?

Karel Lewit, discussing MET methods (Lewit 1985), states that medullary inhibition is not capable of explaining their effectiveness. He considers that the predictable results obtained may relate to the following facts:

- During resistance of minimal force (isometric contraction) only a very few fibres are active, the others being inhibited
- During relaxation (in which the shortened musculature is taken gently to its new limit without stretching) the stretch reflex is

avoided – a reflex which is brought about even by passive and non-painful stretch (see Mattes' views p. 7).

He concludes that this method demonstrates the close connection between tension and pain, and between relaxation and analgesia.

The use of eye movements as part of the methodology is based on research by Gaymans (1980) which indicates, for example, that flexion is enhanced by the patient looking downwards, and extension by looking upwards. Similarly, sidebending is facilitated by looking towards the side involved. These ideas are easily proved, by self-experiment. An attempt to flex the spine, whilst maintaining the eyes in an upwards (towards the forehead) looking position, will be found to be less successful than an attempt made while looking downwards. These eye-direction aids are also useful in manipulation of the joints.

Effects of MET

Lewit discusses the element of passive muscular stretch in MET, and maintains that this factor does not always seem to be essential. In some areas, self-treatment, using gravity as the resistance factor, is effective, and such cases sometimes involve no element of stretch of the muscles in question. Stretching of muscles during MET, according to Lewit, is only required when contracture due to fibrotic change has occurred, and is not necessary if there is simply a disturbance in function. He quotes results in one series of patients in his own clinic in which 351 painful muscle groups, or muscle attachments, were treated by MET (postisometric relaxation) in 244 patients. Analgesia was immediately achieved in 330 cases and there was no effect in only 21 cases. These are remarkable results by any standards.

Lewit suggests, as do many others, that trigger points and 'fibrositic' changes in muscle will often disappear after MET contraction methods. He further suggests that referred local pain points, resulting from problems elsewhere, will also disappear more effectively than where local anaesthesia or needling (acupuncture) methods are employed.

Janda's postfacilitation stretch method

Janda's variation on this approach (Janda 1993), known as 'postfacilitation stretch', uses a different starting position for the contraction and also a far stronger isometric contraction than that suggested by Lewit and most osteopathic users of MET:

1. The shortened muscle is placed in a mid-range position about halfway between a fully stretched and a fully relaxed state.
2. The patient contracts the muscle isometrically, using a maximum degree of effort for 5 to 10 seconds, while the effort is resisted completely.
3. On release of the effort, a rapid stretch is made to a new barrier, without any 'bounce', and this is held for at least 10 seconds.
4. The patient relaxes for approximately 20 seconds and the procedure is repeated between 3 and 5 times more.

Some sensations of warmth and weakness may be anticipated for a short while following this more vigorous approach.

Reciprocal inhibition variation

This method, which forms a component of PNF methodology and MET, is mainly used in acute settings, where tissue damage or pain precludes the use of the more usual agonist contraction, and also commonly as an addition to such methods, often to conclude a series of stretches whatever other forms of MET have been used (Evjenth & Hamberg 1984):

1. The affected muscle is placed in a mid-range position.
2. The patient is asked to push firmly towards the restriction barrier and the operator either completely resists this effort (isometric) or allows a movement towards it (isotonic). Some degree of rotational or diagonal movement may be incorporated into the procedure.
3. On ceasing the effort the patient inhales and exhales fully, at which time the muscle is passively lengthened.

Liebenson notes that 'a resisted isotonic effort towards the barrier is an excellent way in which to facilitate afferent pathways at the conclusion of treatment with active muscular relaxation techniques or an adjustment (joint). This can help reprogram muscle and joint proprioceptors and thus re-educate movement patterns.'

Patient assisted stretch

Aaron Mattes (1990), an innovative researcher into muscle stretching, has noted that for maximum results:

1. The muscle needing stretching must be identified.
2. Precise localisation should be used to ensure that the muscle receives specific stretching.
3. Use should be made of a contractile effort to produce relaxation of the muscles involved.
4. Repetitive isotonic muscle contractions should be used to increase local blood flow and oxygenation.
5. A synchronised breathing rhythm should be established, using inhalation as the part returns to the starting position (the 'rest' phase), and exhalation as the muscle is taken to and through its resistance barrier (the 'work' phase).
6. The muscle to be stretched should be taken into stretch just beyond a point of light irritation – with the patient's assistance – and held for 1 to

2 seconds before being returned to the starting position. Repetitions continue until adequate gain has been achieved.

Mattes uses patient participation in moving the part through the barrier of resistance in order to prevent activation of the myotatic stretch reflex, and this component of his specialised stretching approach has been incorporated into MET methodology by many practitioners.

A variation exists, known as facilitated stretching, in which an acronym (CRAC) is used to describe what is done (Contract–Relax, Antagonist–Contract). As indicated by the words which make up the acronym in this variation, the patient performs all the activity without any passive help (McAtee 1993).

Strengthening variation

Another major variation is to use what has been called isokinetic contraction (also known as progressive resisted exercise). In this the patient, starting with a weak effort but rapidly progressing to a maximal contraction of the affected muscle(s), introduces a degree of resistance to the operator's effort to put the joint, or area, through a full range of motion. The use of isokinetic contraction is reported to be a most effective method of building strength, and to be superior to high repetition, lower resistance exercises (Journal of the American Osteopathic Association 1980). It is also felt that a limited range of motion, with good muscle tone, is preferable (to the patient) to a normal range with limited power. Thus the strengthening of weak musculature in areas of permanent limitation of mobility is seen as an important contribution in which isokinetic contractions may assist.

Isokinetic contractions not only strengthen the fibres which are involved, but have a training effect which enables them to operate in a more coordinated manner. There is often a very rapid increase in strength. Because of neuromuscular recruitment there is a progressively stronger muscular effort as this method is repeated. Isokinetic contractions, and accompanying mobilisation of the region, should take no more than 4 seconds at each contraction, in order to

achieve maximum benefit with as little fatiguing as possible, of either the patient or the operator. Prolonged contractions should be avoided. The simplest, safest, and easiest-to-handle use of isokinetic methods involves small joints, such as those in the extremities. Spinal joints may be more difficult to mobilise while muscular resistance is being fully applied.

The options available in achieving increased strength via these methods, therefore, involve a choice between a partially resisted isotonic contraction, or the overcoming of such a contraction, at the same time as the full range of movement is being introduced (note that both isotonic concentric and eccentric contractions will take place during the isokinetic movement of a joint). Both of these options should involve maximum contraction of the muscles by the patient. Home treatment of such conditions is possible, via self-treatment, as in other MET methods.

Isolytic MET

Another application of the use of isotonic contraction occurs when a direct contraction is resisted and overcome by the operator (Fig. 1.4). This has been termed isolytic contraction, in that

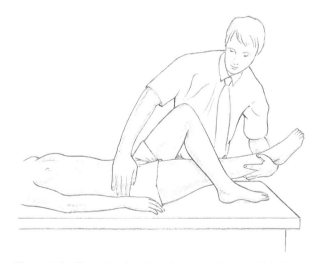

Figure 1.4 Example of an isolytic contraction in which the patient is attempting to move the right leg into abduction towards the right at exactly the same time as the operator is overriding this effort. This stretches the muscles which are contracting (TFL shown in example) thereby inducing a degree of controlled microtrauma, with the aim of increasing the elastic potential of shortened or fibrosed tissues.

it involves the stretching, and sometimes the breaking down, of fibrotic tissue present in the affected muscles. Adhesions of this type are reduced by the application of force by the operator which is just greater than that being exerted by the patient. This procedure can be uncomfortable, and the patient should be advised of this. Limited degrees of effort are therefore called for at the outset of isolytic contractions. This is an isotonic eccentric contraction, in that the origins and insertions of the muscles involved will become further separated, despite the patient's effort to approximate them. In order to achieve the greatest degree of stretch (in the condition of myofascial fibrosis, for example) it is necessary for the largest number of fibres possible to be involved in the isotonic contraction. Thus there is a contradiction in that, in order to achieve this large involvement, the degree of contraction should be a maximal one, and yet this is likely to produce pain, which is contraindicated. It may also, in many instances, be impossible for the operator to overcome.

The patient should be instructed to use about 20% of possible strength on the first contraction, which is resisted and overcome by the operator, in a contraction lasting 3 to 4 seconds. This is then repeated, but with an increased degree of effort on the part of the patient (assuming the first effort was relatively painless). This continuing increase in the amount of force employed in the contracting musculature may be continued until, hopefully, a maximum contraction effort is possible, again to be overcome by the operator. In some muscles, of course, this may require an heroic degree of effort on the part of the operator, and alternative methods are therefore desirable. Deep tissue techniques, such as neuromuscular technique, would seem to offer such an alternative. The isolytic manoeuvre should have as its ultimate aim a fully relaxed muscle, although this will not always be possible.

Why fibrosis occurs naturally

An article in the *Journal of the Royal Society for Medicine* (1983) discusses connective tissue changes:

Aging effects the function of connective tissue more obviously than almost any organ system. Collagen fibrils thicken, and the amounts of soluble polymer decrease. The connective tissue cells tend to decline in number, and die off. Cartilages become less elastic, and their complement of proteoglycans changes both quantitatively and qualitatively. The interesting question is how many of these processes are normal, that contribute blindly and automatically, beyond the point at which they are useful? Does prevention of aging, in connective tissues, simply imply inhibition of crosslinking in collagen fibrils, and a slight stimulation of the production of chondroiten sulphate proteoglycan?

The effects of various soft tissue approaches such as NMT and MET will impact directly on these tissues as well as on the circulation and drainage of the affected structures, which suggests that the aging process can be influenced. Destruction of collagen fibrils, however, is a serious matter (for example when using isolytic stretches), and although the fibrous tissue may be replaced in the process of healing, scar-tissue formation is possible, and this makes repair inferior to the original tissues, both in functional and structural terms. An isolytic contraction has the ability to break down tight, shortened, tissues and the replacement of these with superior material will depend, to a large extent, on the subsequent use of the area (exercise, etc.) as well as the nutritive status of the individual. Collagen formation is dependent on adequate vitamin C, and a plentiful supply of amino acids, such as proline, hydroxyproline and argenine. Manipulation, aimed at the restoration of a degree of normality in connective tissues, should therefore take careful account of nutritional requirements.

The range of choices in stretching, irrespective of the form of prelude to this – strong or mild isometric contraction, starting at or short of the barrier – therefore covers the spectrum from all-passive to all-active, with many variables in between.

PUTTING IT TOGETHER

Many may prefer to use the variations as described above, within individual settings. The recommendation of this text, however, is that they should be 'mixed and matched' so that elements of all of them may be used in any given setting, as appropriate. Lewit's approach seems ideal for more acute and less chronic conditions, while Janda's more vigorous methods seem ideal for hardy patients with chronic muscle shortening.[1]

MET offers a spectrum of approaches which range from those involving hardly any active contraction at all, relying on the extreme gentleness of mild isometric contractions induced by breath-holding and eye movements only, all the way to the other extreme of full-blooded, total-strength contractions. Subsequent to isometric contractions – whether strong or mild – there is an equally sensitive range of choices, involving either energetic stretching or very gentle movement to a new restriction. We can see why Sandra Yale speaks of the usefulness of MET in treating extremely ill patients.

Many patients present with a combination of recent dysfunction (acute in terms of time, if not in degree of pain or dysfunction) overlaid on chronic changes which have set the scene for their acute current problems. It seems perfectly appropriate to use methods which will deal gently with hypertonicity, and more vigorous methods which will help to resolve fibrotic change, in the same patient, at the same time, using different variations on the theme of MET. Other variables can be used which focus on joint restriction, or which utilise RI should conditions be too sensitive to allow PIR methods, or variations on Janda's vigorous stretch methods (Box 1.1).

Discussion of common errors in application of MET will help to clarify these thoughts.

Why MET might be ineffective at times

Poor results from use of MET may relate to an inability to localise muscular effort sufficiently,

since, unless local muscle tension is produced in the precise region of the soft tissue dysfunction, the method is likely to fail to achieve its objectives. Also, of course, underlying pathological changes may have taken place in joints or elsewhere, which make such an approach of short-term value only, since such changes will ensure recurrence of muscular spasms, sometimes almost immediately.

Use of variations such as stretching chronic fibrotic conditions following an isometric contraction and use of the integrated approach (INIT) mentioned earlier in this chapter, represent two examples of further adaptations of Lewit's basic approach which, as described above (p. 5), is ideal for acute situations of spasm and pain.

To stretch or to strengthen?

Marvin Solit, a disciple of Ida Rolf, describes a common error in application of MET – treating the 'wrong' muscles the 'wrong' way (Solit 1963):

As one looks at a patient's protruding abdomen, one might think that the abdominal muscles are weak, and that treatment should be geared towards strengthening them. By palpating the abdomen, however, one would not feel flabby, atonic muscles which would be the evidence of weakness; rather, the muscles are tight, bunched and shortened. This should not be surprising because here is an example of muscle working overtime maintaining body equilibrium. In addition these muscles are supporting the sagging viscera, which normally would be supported by their individual ligaments. As the abdominal muscles are freed and lengthened, there is a general elevation of the rib cage, which in turn elevates the head and neck.

Attention to tightening and hardening these supposedly weak muscles via exercise, observes Solit, results in no improvement in posture, and no reduction in the 'pot-bellied' appearance. Rather, the effect is to further depress the thoracic structures, since the attachments of the abdominal muscles, superiorly, are largely onto the relatively mobile, and unstable, bones of the rib cage. Shortening these muscles simply achieves a degree of pull on these structures, towards the stable pelvic attachments below. The approach to this problem, adopted by

---

[1]The term 'acute' is used here to indicate problems which have arisen over the previous 3 weeks or less, or which are acutely sensitive. Any other soft-tissue dysfunction is regarded as chronic and is assumed to contain some degree of fibrotic change, calling for stretching subsequent to MET contractions of one sort or another (agonist if possible, antagonist if not).

Box 1.1   Defining the terms used in MET

The terms used in MET require clear definition and emphasis:

1. An isometric contraction is one in which a muscle, or group of muscles, or a joint, or region of the body, is called upon to contract, or move in a specified direction, and in which that effort is matched by the operator's effort, so that no movement is allowed to take place.
2. An isotonic contraction is one in which movement does take place, in that the counterforce offered by the operator is either less than that of the patient, or is greater.

In the first isotonic example there would be an approximation of the origin and insertion of the muscle(s) involved, as the effort exerted by the patient more than matches that of the operator. This has a tonic effect on the muscle(s) and is called a concentric isotonic contraction. This method is useful in toning weakened musculature.

The other form of isotonic contraction involves an eccentric movement in which the muscle, while contracting, is stretched. The effect of the operator offering greater counterforce than the patient's muscular effort is to lengthen a muscle which is trying to shorten. This is also called an isolytic contraction. This manoeuvre is useful in cases where there exists a marked degree of fibrotic change. The effect is to stretch and alter these tissues – inducing controlled microtrauma – thus allowing an improvement in elasticity and circulation.

Direct and indirect

It is sometimes easier to describe these variations simply in terms of whether the operator's force is the same as, less than, or greater than that of the patient. In any given case there is going to exist a degree of limitation in movement, in one direction or another, which may involve purely soft-tissue components of the area, or actual joint restriction (in such cases there is bound to be some involvement of soft tissues). The operator establishes, by palpation and by mobility assessments (motion palpation), the direction of maximum 'bind' or restriction. This is felt as a definite point of limitation in one or more directions. In many instances the muscle(s) will be shortened and incapable of stretching and relaxing. Should the isometric, or isotonic, contraction which the patient is asked to perform, be one in which the contraction of the muscles or movement of the joint is away from the barrier or point of bind, whilst the operator is using force in the direction which goes towards, or

through that barrier, then this form of treatment involves what is called a direct action. Should the opposite apply, with the patient attempting to take the area/joint/muscle towards the barrier, while the operator is resisting, then this is an indirect manoeuvre.

Experts differ

As with so much in manipulative terminology, there is disagreement even in this apparently simple matter of which method should be termed 'direct' and which 'indirect'. Grieve (1984) describes the variations thus: 'Direct action techniques in which the patient attempts to produce movement towards, into or across a motion barrier; and indirect techniques, in which the patient attempts to produce motion away from the motion barrier, i.e. the movement limitation is attacked indirectly.'

On the other hand, Goodridge (1981), having previously illustrated and described a technique where the patient's effort was directed away from the barrier of restriction, states: 'The aforementioned illustrations used the direct method. With the indirect method the component is moved by the operator away from the restrictive barrier.'

If the operator is moving away from the barrier, then the patient is moving towards it, and in Goodridge's terminology (i.e. osteopathic) this is an indirect approach. In Grieve's terminology (physiotherapy) this is a direct approach. Plainly these views are contradictory.

David Heiling, of the Philadelphia College of Osteopathic Medicine, notes that, since muscle energy techniques always involve two opposing forces (the patient's and the operator's), he feels that it is more logical to indicate which force is being used in order to characterise a given technique. Thus an operator-direct method is also equally accurately described as a patient-indirect method. Heiling feels that operator-direct methods (in which the patient is using muscles usually already in spasm, or shortened, and therefore sometimes described as 'using the agonist') are more appropriate to managing chronic conditions, rather than acute ones. When acute, the shortened muscles could have oedema or fibre damage, and could go further into spasm were they employed, and it would therefore be safer to contract their antagonists.

Operator-direct methods are particularly suitable for rehabilitation where muscle shortening has occurred. Patient-direct techniques are more suitable for acute conditions, where the antagonists to the shortened muscles are called on to contract.

Rolfers, is to free and loosen these overworked, and only apparently weakened, tissues. This allows for a return to some degree of normality, freeing the tethered thoracic structures, and thus correcting the postural imbalance. Attention to the shortened, tight musculature, which will also be inhibiting their antagonist muscles,

should be the primary aim. Exercise is not suitable at the outset, before this primary goal is achieved (Fig. 1.5).

The common tendency, in some schools of therapy, to encourage the strengthening of weakened muscle groups in order to normalise postural and functional problems, is also

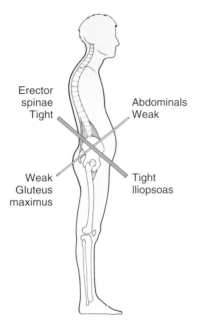

Erector
spinae
Tight

Abdominals
Weak

Weak
Gluteus
maximus

Tight
Iliopsoas

Figure 1.5 Lower crossed syndrome. An example of a
common postural imbalance pattern, involving a chain
reaction of hypertonia and hypotonia in which excessively
tight and short muscles are inhibiting their antagonists.

discussed by Vladimir Janda (1978). He expresses the reasons why this approach is literally putting the cart before the horse: 'In pathogenesis, as well as in treatment of muscle imbalance and back problems, tight muscles play a more important, and perhaps even primary, role in comparison to weak muscles.' He continues with the following observation:

Clinical experience, and especially therapeutic results, support the assumption that (according to Sherrington's law of reciprocal innervation) tight muscles act in an inhibitory way on their antagonists. Therefore, it does not seem reasonable to start with strengthening of the weakened muscles, as most exercise programmes do. It has been clinically proved that it is better to stretch tight muscles first. It is not exceptional that, after stretching of the tight muscles, the strength of the weakened antagonists improves spontaneously, sometimes immediately, sometimes within a few days, without any additional treatment.

This sound, well-reasoned, clinical and scientific observation, which directs our attention and efforts towards the stretching and normalising of those tissues which have shortened and

tightened seems irrefutable, and this theme will be pursued further in Chapter 2.

MET is designed to assist in this endeavour and, as discussed above, also provides an excellent method for assisting in the toning of weak musculature, should this still be required, after the stretching of the shortened antagonists, by use of isotonic methods.

Tendons

Aspects of the physiology of muscles and tendons are worthy of a degree of review, insofar as MET and its effects are concerned. The tone of muscle is largely the job of the Golgi tendon organs. These detect the load applied to the tendon, via muscular contraction. Reflex effects, in the appropriate muscles, are the result of this information being passed from the Golgi tendon organ back along the cord. The reflex is an inhibitory one, and thus differs from the muscle spindle stretch reflex. Sandler (1983) describes some of the processes involved:

When the tension on the muscles, and hence the tendon, becomes extreme, the inhibitory effect from the tendon organ can be so great that there is sudden relaxation of the entire muscle under stretch. This effect is called the lengthening reaction, and is probably a protective reaction to the force which, if unprotected, can tear the tendon from its bony attachments. Since the Golgi tendon organs, unlike the [muscle] spindles, are in series with the muscle fibres, they are stimulated by both passive and active contractions of the muscles.

Pointing out that muscles can either contract with constant length and varied tone (isometrically), or with constant tone and varied length (isotonically), he continues: 'In the same way as the gamma efferent system operates as a feedback to control the length of muscle fibres, the tendon reflex serves as a reflex to control the muscle tone.'

The relevance of this to soft tissue techniques is explained thus:

In terms of longitudinal soft tissue massage, these organs are very interesting indeed, and it is perhaps the reason why articulation of a joint, passively, to stretch the tendons that pass over the joint, is often as effective in relaxing the soft tissues as direct massage of the muscles themselves. Indeed, in some cases, where the muscle is actively in spasm, and is likely to object to being pummelled directly, articulation, muscle energy technique, or functional balance techniques, that make use of the tendon organ reflexes, can be most effective.

The use of this knowledge in therapy is obvious and Sandler explains part of the effect of massage on muscle as follows: 'The [muscle] spindle and its reflex connections constitute a feedback device which can operate to maintain constant muscle length, as in posture; if the muscle is stretched the spindle discharges increase, but if the muscle is shortened, without a change in the rate of gamma discharge, then the spindle discharge will decrease, and the muscle will relax.' He believes that massage techniques cause a decrease in the sensitivity of the gamma efferent, and thus increase the length of the muscle fibres, rather than a further shortening of them; this produces the desired relaxation of the muscle. MET provides for the ability to influence both the muscle spindles, and also the Golgi tendon organs.

Joints and MET

Bourdillon (1982) tells us that shortening of muscle seems to be a self-perpetuating phenomenon, which results from an over-reaction of the gamma-neurone system. It seems that the muscle is incapable of returning to a normal resting length as long as this continues. Whilst the effective length of the muscle is thus shortened, it is nevertheless capable of shortening further. The pain factor seems related to the muscle's inability thereafter, to be restored to its anatomically desirable length. The conclusion is that much joint restriction is a result of muscular tightness and shortening. The opposite may also apply, where damage to the soft or hard tissues of a joint is a factor. In such cases the periarticular and osteophytic changes, all too apparent in degenerative conditions, are the major limiting factor in joint restrictions. In both situations, however, MET may be useful, although

more useful where muscle shortening is the primary factor.

The restriction which takes place as a result of tight, shortened muscles is usually accompanied by some degree of lengthening and weakening of the antagonists. A wide variety of possible permutations exists, in any given condition, involving muscular shortening, which may be initiating, or be secondary to, joint dysfunction, combined with weakness of antagonists. A combination of isometric and isotonic methods can effectively be employed to lengthen and stretch the shortened groups, and to strengthen and shorten the weak, overlong muscles.

Paul Williams (1965) states a basic truism which is often neglected by the professions which deal with musculoskeletal dysfunction when he says:

The health of any joint is dependent upon a balance in the strength of its opposing muscles. If for any reason a flexor group loses part, or all of its function, its opposing 'tensor group will draw the joint into a hyperextended position, with abnormal stress on the joint margins. This situation exists in the lumbar spine of modern man.

Lack of attention to the muscular component of joints in general, and spinal joints in particular, results in frequent inappropriate treatment of the joints thus affected. Correct understanding of the role of the supporting musculature would frequently lead to normalisation of these tissues, without the need for heroic manipulative efforts. MET and other soft tissue approaches focus attention on these structures, and offer the opportunity to correct both the weakened musculature, and the shortened, often fibrotic, antagonists.

Several studies will be detailed (Chapters 4 and 6), showing the effectiveness of MET application in diverse population groups, including a Polish study on the benefits of MET in joints damaged by haemophilia, and a Swedish study on the effects of MET in treating lumbar spine dysfunction, as well as an American/Czech study involving myofascial pain problems. In the main, the results indicate a universal role in providing resolution or relief for this safe and effective approach.

REFERENCES

Bourdillon J 1982 Spinal manipulation, 3rd edn. Heinemann, London

Blood S 1980 Treatment of the sprained ankle. Journal of the American Osteopathic Association 79(11): 689

Chaitow L 1993 Integrated Neuromuscular Inhibition Technique (INIT) in treatment of pain and trigger points. British Journal of Osteopathy XIII: 17–21

DiGiovanna E 1991 Osteopathic approach to diagnosis and treatment. Lipincott, Philadelphia

Evjenth O, Hamberg J 1984 Muscle stretching in manual therapy. Alfta, Sweden

Gaymans F 1980 Die Bedueting der atemtypen fur mobilisation der werbelsaule maanuelle. Medizin 18: 96

Goodridge J P 1981 Muscle energy technique: definition, explanation, methods of procedure. Journal of the American Osteopathic Association 81(4): 249–254

Greenman P 1989 Principles of manual medicine. Williams and Wilkins, Baltimore

Grieve G P 1984 Mobilisation of the spine. Churchill Livingstone, Edinburgh, p 190

Janda V 1978 In: Korr I (ed) Neurobiological mechanisms in manipulative therapy. Plenum Press, New York

Janda V 1989 Muscle function testing. Butterworths, London

Janda V 1993 Presentation to Physical Medicine Research Foundation, Montreal Oct 9–11

Journal of the Royal Society for Medicine 1983 Connective tissues: the natural fibre reinforced composite material. Journal of the Royal Society for Medicine 76

Kabot H 1950 Studies of neuromuscular dysfunction. Kaiser Permanente Foundation Medial Bulletin 8: 121–143

Levine M et al 1954 Relaxation of spasticity by physiological techniques. Archives of Physical Medicine 35: 214–223

Lewit K, Simons D 1984 Myofascial pain: relief by post isometric relaxation. Archives of Physical Medical Rehabilitation 65: 452–456

Lewit K 1985 Manipulative therapy in rehabilitation of the motor system. Butterworths, London

Lewit K 1986a Muscular patterns in thoraco-lumbar lesions. Manual Medicine 2, p 105

Lewit K 1986b Postisometric relaxation in combination with other methods of muscular facilitation and inhibition. Manual Medicine 2: 101–104

Lewit K 1991 Manipulative therapy in rehabilitation of the locomotor system – expanded version. Butterworths, London

Liebenson C 1989 Active muscular relaxation techniques (part 1). Journal of Manipulative and Physiological Therapeutics 12(6): 446–451

Liebenson C 1990 Active muscular relaxation techniques (part 2). Journal of Manipulative and Physiological Therapeutics 13(1): 2–6

McAtee R 1993 Facilitated stretching. In: Human Kinetics. Champaign, Illinois

Mattes A 1990 Flexibility – active and assisted stretching. Mattes, Sarasota

Mitchell F L Snr. 1958 Structural pelvic function. Yearbook of the Academy of Osteopathy 1958 (expanded in references in 1967 yearbook), p 71

Mitchell F L Snr. 1967 Motion discordance. Yearbook of the Academy of Applied Osteopathy, Carmel 1967: 1–5

Mitchell F Jr., Moran P S, Pruzzo N 1979 An evaluation and treatment manual of osteopathic muscle energy procedures. Valley Park, Illinois

Moritan T et al 1987 Activity of the motor unit during concentric and eccentric contractions. American Journal of Physiology 66: 338–350

Sandler S 1983 Physiology of soft tissue massage. British Osteopathic Journal 15: 1–6

Solit M 1963 A study in structural dynamics. Yearbook of Academy of Applied Osteopathy 1963

Stiles E 1984a Patient Care May 15: 16–97

Stiles E 1984b Patient Care Aug 15: 117–164

Travell J, Simons D 1992 Myofascial pain and dysfunction, Volume 2. Williams and Wilkins, Baltimore

Williams P 1965 The lumbo-sacral spine. McGraw Hill, New York

2

Patterns of function and dysfunction

ASSESSING SYMPTOMS

It is necessary to examine the viewpoints of different experts if we are to come to an understanding of soft tissue dysfunction in particular, and of its place in the larger scheme of things in relation to musculoskeletal and general dysfunction. A commonality will be noted in many of the views which will be presented, as well as distinctive differences in emphasis. It is not the position of the authors to be dogmatic, but to present evidence from which the reader can make choices.

Most models include a progression, a sequence of events, a chain reaction, and a process of adaptation, modification, attempted homeostatic accommodation to whatever is taking place.

In order to adequately deal with soft tissue or joint dysfunction, it is axiomatic that what is dysfunctional should first be accurately assessed and identified. Based on such verifiable data as are available, a treatment plan with a realistic prognosis can be formulated, irrespective of the methods of treatment chosen. The assessment findings are then capable of being used as a yardstick against which results can be assessed and evaluated. If progress is not forthcoming, a reassessment is required.

Among the many pertinent questions which need answering are :

1. Which muscle groups have shortened and contracted?

2. Is the evident restriction in a specific soft tissue structure related to neuromuscular influence (which could be recorded on an EMG reading of the muscle), or tightness due to connective tissue fibrosis (which would not show on an EMG reading), or both?

3. Which muscles have become significantly weaker, and is this through inhibition or through atrophy?

4. What 'chain reactions' of functional imbalance have occurred as one muscle group (because of its excessive hypertonicity) has inhibited and weakened its antagonists?

5. What joint restrictions are associated with these soft tissue changes – either as a result, or as a cause of these?

6. Is a restriction primarily of soft tissue or of joint origin, or a mixture of both?

7. How does the obvious dysfunction relate to the nervous system and to the rest of the musculoskeletal system of this patient?

8. What patterns of compensating postural stress have such changes produced (or have produced them) and how is this further stressing the body as a whole, affecting its energy levels and function?

9. Within particular muscle areas which are stressed, what local soft tissue changes (fascia etc.) have occurred leading, for example, to myofascial trigger point development?

10. What symptoms, whether of pain or other forms of dysfunction, are the result of reflexogenic activity such as trigger points?

In other words, what palpable, measurable, identifiable evidence is there which connects what we can observe, test and palpate to the symptoms (pain, restriction, fatigue etc.) of this patient?

And further:

11. What, if anything, can be done to remedy or modify the situation, safely and effectively?

12. Is this a self-limiting condition which treatment can make more tolerable as it normalises?

13. Is this a condition which can be helped towards normalisation by therapeutic intervention?

14. Is this a condition which cannot normalise but which can be modified to some extent, so making function easier or reducing pain?

15. What mobilisation, relaxation and/or strengthening strategies are most likely to be of assistance, and how can this patient learn to use themselves less stressfully following treatment?

16. To what degree can the patient participate in the process of recovery, normalisation, rehabilitation?

Fortunately, as a part of such therapeutic intervention, a vast range of muscle energy techniques exist which can be taught as self-treatment, thus involving and empowering the patient.

VIEWING SYMPTOMS IN CONTEXT

Clearly the answers to this range of questions will vary enormously from person to person, even if symptoms appear similar at the outset. The context within which symptoms appear and exist will largely determine the opportunities available for successful therapeutic intervention.

Pain is probably the single most common symptom experienced by humans and, along with fatigue, is the most frequent reason for anyone consulting a doctor in industrialised societies – indeed the World Health Organisation has suggested that pain is 'the primary problem' for developed countries (WHO 1981).

Within that vast area of pain, musculoskeletal dysfunction in general, and back pain in particular, feature large. If symptoms of pain and restriction are viewed in isolation, with inadequate attention being paid to the degree of acuteness or chronicity, their relationship with the whole body and its systems (including the musculoskeletal and nervous systems) – as well as, for example, the emotional and nutritional status of the individual and of the multiple environmental, occupational, social and other factors which impinge upon them – then it is quite possible that they will be treated inappropriately.

A patient with major social, economic and emotional stressors current in her life and who presents with muscular pain and backache, is unlikely to respond – other than in the short term – to manual approaches which ignore the enormous and multiple coping strain she is handling. In many instances, the provision of a job, a new home, a new spouse (or removal of the present one) would be the most appropriate treatment in terms of addressing the real causes of such pain or backache. However, the manual therapist/practitioner must utilise those skills available, so that suitable treatment will, if nothing else, minimise the patient's mechanical and functional strains – even if they do not always deal with what is really wrong!

Suitable treatment for pain and dysfunction which has evolved out of the somatisation by the patient of emotional distress, might well be helped more through application of deep relaxation methods, non-specific 'wellness' bodywork methods and/or counselling and enhancement of stress-coping abilities, rather than specific musculoskeletal interventions which impose yet another adaptation demand on an already overextended system. The art of successfully applied manual approaches to healing lies, at least in part, in recognising when intervention should be specific and when it needs to be more general.

The role of the emotions in musculoskeletal dysfunction

Sandman has analysed the interaction between mind influences on those neurological and metabolic functions which regulate physiological responses, and concludes that there is a synergistic relationship which results in a need to address both the psychological and physiological aspects of stress which have emerged from the effects of (among others) familial, relationship, career, social, health, traumatic and financial stressors (Sandman 1984, Selye 1976). Unless both aspects (mind and body) are addressed, 'no permanent reduction of the negative feedback loop is possible'.

Sandman reviews the process by means of which stress and secondary stress influence muscles:

1. Stress causes biochemical changes in the brain – partly involving neurotransmitter production which increases neural excitability.
2. Postural changes follow in muscles, commonly involving increased tone which retards circulatory efficiency and increases calcium, lactic acid and hyaluronic acid accumulation.
3. Local contractile activity in muscle is increased because of the interaction between calcium and adenosine triphosphate (ATP),

leading to physiological contractions which shorten and tense muscle bundles.
4. Sustained metabolic activity in such muscles increases neural hyperreactivity which may stimulate reflex vasoconstriction, leading to local tenderness and referred pain.
5. Relative oxygen lack and reduced energy supply results from decreased blood flow leading to an energy-deficient muscle contraction in which the sarcoplasmic reticulum becomes damaged.
6. The energy-sensitive calcium pump responds by increasing muscle contraction due to the lack of energy supply, leading to ever greater depletion.
7. Pain is a feature of this process, possibly due to accumulation locally of chemicals, which might include bradykinin, Substance-P, inflammatory exudates, histamine and others.
8. Local pressure build-up, involving these chemicals and local metabolic wastes, and/or local ischaemia, are sufficient causes to produce local spasm which might involve local and/or referred pain.
9. If, at this time, the muscle is stretched, the locked actin and myosin filaments will release the contraction and sufficient ATP can then accumulate to allow a more normal sarcoplasmic reticulum, which would allow for removal of the build-up of metabolites.
10. The degree of damage which the muscle sustains due to this sequence depends entirely upon the length of time during which these conditions are allowed to continue: 'At this point physiological aspects as well as psychological should be addressed ... to stop the debilitating cycle.'

Sandman's method of relieving the physical aspects of the condition involves active and passive stretching alongside pressure and vibratory techniques.

Korr's 'orchestrated movement' concept

It is necessary to conceptualise muscular function and dysfunction as being something other than a local event. Irwin Korr (1976) stated the position elegantly and eloquently:

The spinal cord is the keyboard on which the brain plays when it calls for activity. But each 'key' in the console sounds not an individual 'tone' such as the contraction of a particular group of muscle fibres, but a whole 'symphony' of motion. In other words, built into the cord is a large repertoire of patterns of activity, each involving the complex, harmonious, delicately balanced orchestration of the contractions and relaxation of many muscles. The brain thinks in terms of whole motions, not individual muscles. It calls, selectively, for the preprogrammed patterns in the cord and brain stem, modifying them in countless ways and combining them in an infinite variety in still more complex patterns. Each activity is subject to further modulation refinement, and adjustment by the feedback continually streaming in from the participating muscles, tendons and joints.

We must never forget the complex inter-relationships between the soft tissues, the muscles, fascia and tendons and their armies of neural reporting stations, as we attempt to understand the nature of dysfunction and of what is required to achieve normalisation.

A proprioceptive model of dysfunction

Let us visualise an area at relative ease, in which there is some degree of difference between antagonist muscles, one group comfortably stretched, the other short of their normal resting length, and equally comfortable, such as might exist in someone comfortably bending forwards to lift something. Imagine a sudden demand for stability in this setting. As this happened, the annulospiral receptors in the short (flexor) muscles would respond to the sudden demand (the person or whatever they are lifting un-accountably slips for example) by contracting even more (Mathews 1981).

The neural reporting stations in these shortened muscles (which would be rapidly changing length to provide stability) would be firing impulses as if the muscles were being stretched, even when the muscle remained well short of its normal resting length. At the same time the stretched extensor muscles would rapidly shorten in order to stabilise the situation. Once stability has been achieved, they are likely to still be somewhat longer than their normal resting length.

Korr (1947, 1975) has described what happens in the abdominal muscles (flexors) in such a situation. He says that because of their relaxed status, short of their resting length, there occurs a silencing of the spindles. However, due to the demand for information from the higher centres, gamma gain is increased reflexively and as the muscle contracts rapidly to stabilise the alarm demands, the central nervous system would receive information that the muscle which is actually short of its neutral resting length was being stretched. In effect, the muscles would have adopted a position of somatic dysfunction as a result of 'garbled' or inappropriate proprioceptive reporting. As DiGiovanna (1991) explains,

With trauma or muscle effort against a sudden change in resistance, or with muscle strain incurred by resisting the effects of gravity for a period of time, one muscle at a joint is strained and its antagonist is hyper-shortened. When the shortened muscle is suddenly stretched the annulospiral receptors in that muscle are stimulated causing a reflex contraction of the already shortened muscle. The proprioceptors in the short muscle now fire impulses as if the shortened muscle were being stretched. Since this inappropriate proprioceptor response can be maintained indefinitely a somatic dysfunction has been created.

In effect, the two opposing sets of muscles would have adopted a stabilising posture to protect the threatened structures, and in doing so would have become locked into positions of imbalance in relation to their normal function. One would be shorter and one longer than their normal resting length. At this time any attempt to extend the area/joint(s) would be strongly resisted by the tonically shortened flexor group. The individual would be locked into a forward bending distortion (in our example). The joint(s) involved would not have been taken beyond their normal physiological range and yet the normal range would be unavailable due to the shortened status of the flexor group (in this particular example). Going further into flexion, however, would present no problems or pain.

Walther (1988) summarises the situation as follows (Fig. 2.1):

When proprioceptors send conflicting information there may be simultaneous contraction of the antagonists . . . without antagonist muscle inhibition,

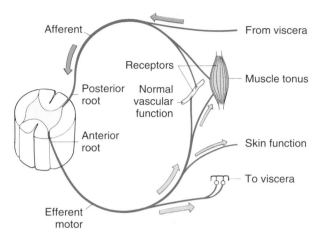

Figure 2.1A   Schematic representation of normal afferent influences deriving from visceral, muscular and venous sources, on the efferent supply to those same structures.

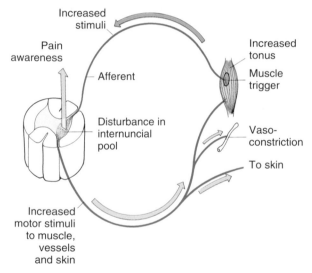

Figure 2.1B   Schematic representation of normal afferent influences deriving from a muscle which displays excessively increased tonus and/or trigger point activity, both in pain awareness and on the efferent motor supply to associated muscular, venous and skin areas.

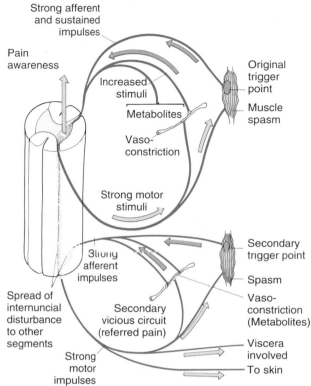

Figure 2.1C   Schematic representation of the secondary spread of neurologically induced influences deriving from acute or chronic soft tissue dysfunction, and involving trigger point activity and/or spasm.

joint and other strain results . . . a reflex pattern develops which causes muscle or other tissue to maintain this continuing strain. It [strain dysfunction] often relates to the inappropriate signalling from muscle proprioceptors that have been strained from rapid change that does not allow proper adaptation.

We can recognise this 'strain' situation in an acute setting in torticollis, in whiplash, as well as in acute 'lumbago'. It is also recognisable as a feature of many types of chronic somatic dysfunction in which joints remain restricted due to muscular imbalances of this type.

Van Buskirk's nociceptive model

A variation on the theme of a progression of dysfunctional changes has been proposed by Van Buskirk (1990) who suggests the following sequence:

1. Nociceptors (peripheral pain receptors) in a muscle are activated by minor trauma from a chemical, mechanical, thermal or other damaging stimuli source (any disease or trauma in any somatic or visceral structure produces nociceptive activation).

2. Nociceptive activation transmits impulses to other axons in the same nociceptor as well as to the spinal cord.

3. Various peptide transmitters in the axon branches are released resulting in vasodilatation and the gathering of immune cells around and in the trauma site.

4. These in turn release chemicals, which enhances the vasodilatation and extravasation while also lowering the nociceptive threshold.

5. Organs at a distance may display axon reflex effects; for example, skeletal muscles and the heart may be simultaneously affected.

6. Spinal neurons will be stimulated by impulses entering the cord synaptically, which influences aspects of the higher CNS which registers pain; or the impulses might stimulate preganglionic autonomic neurons or even the spinal skeletal muscle motor pool, producing nocifensive reflexes.

7. There may be poor localisation of pain at this stage, if it is perceived at all, due to the many sources influencing the same spinal neurons, as well as the divergence of signals along neighbouring spinal segments. Pain will, however, be most noticeable in the originating segment.

8. Any sympathetic response to this chain of events will depend upon the effects of sympathetic stimulus to the target organ, and could (among others) involve cardiopressor, gastrointestinal stasis, bronchodilatation, vasopressor or vasodilator or negative immune function effects.

9. Muscular responses could involve local or multisegmental changes, including shortening of the injured muscle itself via synergistic or self-generated action from non-injured fibres; or overlying muscle might attempt to guard underlying tissues, or some other defensive action might ensue.

10. Direct mechanical restriction of the affected muscles derives from vasodilatation which, along with chemicals associated with tissue injury – bradykinin, histamine, serotonin etc. – causes stimulation of local nociceptors in the muscle associated with the original trauma, or those reflexively influenced.

11. A new defensive muscular arrangement will develop which will cause imbalance and a shortening of the muscles involved. These will not be held at their maximal degree of shortening nor in their previously neutral position.

12. This continued contraction results in additional nociceptive action as well as fatigue, which tends to cause recruitment of additional muscular tissues to maintain the abnormal situation.

13. After a matter of hours or days the abnormal joint positions which result from this defensive muscular activity become chronic as connective tissue reorganisation involving tissue fibrocytes commences.

14. Connective tissues will be randomly orientated in the shortened muscles and less capable of handling stress along normal lines of force.

Van Buskirk describes the progression as follows: 'In the lengthened muscles, creep will elongate the connective tissue, producing slack without stressing the lengthened muscles. Now maintenance of the joint in the non-neutral position dictated by both the nocifensive reflexes and the connective tissue changes no longer requires continuous muscle activity.' Now:

- Active contraction only occurs when the area is stressed which would reactivate the nociceptors.
- At the same time the joint is neither 'gravitationally, posturally, nor functionally balanced' making it far more likely to be stressed and to produce yet more nociceptive activation.
- There would now exist a situation of restricted motion deriving from the original shortening, chronic nociceptive activation, as well as autonomic activation.

In effect, there is now neurologically-derived restriction as well as structural modifications, fibrotic connective tissue changes, both of which require normalising in order to restore normal function. Both the original tissues which were stressed as well as others which have modified in a protective manner are influencing the unbalanced, unphysiological situation.

An example of nociceptive modulated dysfunction. Let us consider someone involved in a simple neck stress as their car came to an unexpected halt. The neck would be thrown backwards into hyperextension, stressing the flexor group of muscles. The extensor group would be rapidly shortened and various proprioceptive changes leading to strain and reflexive shortening would operate (as described above in relation to a bending strain) inducing them to remain in a shortened state. At the time of the sudden hyperextension, the flexors of the neck would be violently stretched inducing actual tissue damage.

Nociceptive responses (which are more powerful than proprioceptive influences) would occur and these multisegmental reflexes would produce a flexor withdrawal – increasing tone in the flexor muscles.

The neck would now have hypertonicity of both the extensors and the flexors – pain, guarding and stiffness would be apparent and the role of clinician would be to remove these restricting influences layer by layer.

Where pain is a factor in strain this has to be considered as producing an overriding influence over whatever other more 'normal' (proprioceptive) reflexes might be operating. In the example of neck strain described, it is obvious that in real life matters are likely to be even more complicated since a true whiplash would introduce both rapid hyperextension and hyperflexion, so producing a multitude of conflicting layers of dysfunction.

The proprioceptive and nociceptive reflexes which might be involved in the production of strain are likely to also involve other factors. As Bailey (Bailey & Dick 1992) explains: 'Probably few dysfunctional states result from a purely proprioceptive or nociceptive response. Additional factors such as autonomic responses, other reflexive activities, joint receptor responses, or emotional states must also be accounted for.'

However, it is at the level of our basic neurological awareness that understanding of the complexity of these problems commences and we need to be aware of the choices which are available for resolving such dysfunction.

How would MET be able to influence this situation? Various approaches are likely to be helpful, including a variety of techniques derived from positional release methods, such as Strain/

CounterStrain (SCS) (Jones 1964), Facilitated Positional Release (DiGiovanna 1991), Functional Technique (Greenman 1989) etc., as well as various modifications of muscle energy technique.

Van Buskirk states it thus:

In indirect 'muscle energy' the skeletal muscles in the shortened area are initially stretched to the maximum extent allowed by the somatic dysfunction [to the barrier]. With the tissues held in this position the patient is instructed to contract the affected muscle voluntarily. This isometric activation of the muscle will stretch the internal connective tissues. Voluntary activation of the motor neurons to the same muscles also blocks transmission in spinal nociceptive pathways. Immediately following the isometric phase, passive extrinsic stretch is imposed, further lengthening the tissues towards the normal easy neutral position.

It is as well to emphasise that these models of the possible chain reaction of events taking place in acute and chronic musculoskeletal dysfunction are included in order to help us to understand what might be happening in the complex series of events which surround, and which flow from, such problems. These elegant attempts at interpreting our understanding of stress and strain are not definitive; there are other models, and some of them will be touched on as we progress through our exploration of the patterns of dysfunction which confront us clinically.

Janda's 'primary and secondary' responses

It has become a truism that we need to consider the body as a whole. However, local focus still seems to be the dominant clinical approach. Janda (1988) gives examples of why this is shortsighted in the extreme.

He discusses the events which follow on from the presence of a short leg – which might well include an altered pelvic position, scoliosis, altered head position, changes at the cervicocranial junction, compensatory activity of the small cervico-occipital muscles, later compensation of neck musculature, increased muscle tone, muscle spasm, probable joint dysfunction, particularly at cervicocranial junction . . . and a sequence of events which would then include compensation and adaptation responses in many muscles, followed by the evolution of a variety of possible syndromes involving head/neck, TMJ, shoulder/arm or others (see discussion of upper and lower 'crossed' syndromes later in this chapter, pp 32–33).

Janda's point is that at such a time, after all the adaptation that has taken place, treatment of the most obvious cervical restrictions, where the patient might be aware of pain and restriction, would offer limited benefit.

He points to the existence of oculopelvic and pelviocular reflexes which indicate that any change in pelvic orientation alters the position of the eyes and vice versa, and to the fact that eye position modifies muscle tone, particularly the suboccipital muscles (look up and extensors tighten, look down and flexors prepare for activity etc.). The implications of modified eye position due to altered pelvic position therefore becomes yet another factor to be considered as we try to unravel chain reactions of interacting elements (Komendatov 1945).'These examples,' Janda says, 'serve to emphasise that one should not limit consideration to local clinical symptomatology . . . but [that we] should always maintain a general view.'

Isaacson's 'functional unit'

Isaacson (1980) helps us to understand the interaction of associated parts in terms of spinal motion. He describes spinal muscles as being divided into two groups, with one set being prime movers (extrinsic) and the others stabilisers (intrinsic) including the erector spinae muscle mass. Although the component parts of the erector spinae muscle group are often referred to individually, as discrete entities – for example multifidus, intertransverse, interspinal etc. – this is basically inaccurate. He states that, 'various functions have been assigned to these intrinsic muscles, on the assumption that they actually move vertebrae; however, the arrangement and position of the muscle bundles, making up this group, would seem to make it improbable that they have much to do in this regard.' They are, instead, stabilisers and proprioceptive sensory receptors which facilitate the coordinated activity of the vertebral complex (as in Korr's 'whole motions').

The force required to move the vertebral column comes from the large, extrinsic, muscles. Analysis of the multifidus group, which is particularly thick in the lumbar region, indicates that its component fascicles could not be prime movers, and that they serve effectively as maintainers of the position, normal or abnormal, in which the prime movers place the vertebrae. The same finding is made in relation to the semispinal group of muscles. These are responsible for compensatory lesions, derived from the vertebra above and below, by virtue of the arrangement of groups of pairs of stabilising fascicles. These groups of muscles are, Isaacson maintains, responsible in large part for the co-ordinated, synchronous, function of the spinal column which is a complex of the two functions of the different types of muscles in the region; those that stabilise, and those that move. Isaacson goes so far as to suggest that the evidence points to the spinal region being a vast network of information gathering tissues: 'Arranged as they are in a variety of positions some of the individual muscle bundles are placed on a stretch by any change of position in the vertebral column, and the tension so produced is translated into terms of proprioceptive sensation and reported to the CNS.'

Thus the vertebral column and the body must be viewed as a functional unit, and not as a collection of parts and organs which function independently of each other. This is a concept which, while obvious, is often neglected in practice.

As we will discover later in this chapter (p. 28) not only do extrinsic prime movers and intrinsic stabilisers behave differently in their normal function but also, most importantly, in their dysfunction.

Fascial considerations

If we are to have anything like a clear overview of soft tissue dysfunction it is necessary to add into the equation the influence of fascia which invests, supports, divides, enwraps, gives cohesion to and is an integral part of every aspect of soft tissue structure and function throughout the body and which represents a single structural entity, from the inside of the skull to the soles of the feet.

Rolf (1962) puts fascia and its importance into perspective when she discusses its properties:

Our ignorance of the role of fascia is profound. Therefore even in theory it is easy to overlook the possibility that far-reaching changes may be made not only in structural contour, but also in functional manifestation, through better organisation of the layer of superficial fascia which enwraps the body. Experiment demonstrates that drastic changes may be made in the body, solely by stretching, separating and relaxing superficial fascia in an appropriate manner. Osteopathic manipulators have observed and recorded the extent to which all degenerative changes in the body, be they muscular, nervous, circulatory or organic, reflect in superficial fascia. Any degree of degeneration, however minor, changes the bulk of the fascia, modifies its thickness and draws it into ridges in areas overlying deeper tensions and rigidities. Conversely, as this elastic envelope is stretched, manipulative mechanical energy is added to it, and the fascial colloid becomes more 'sol' and less 'gel'. As a result of the added energy, as well as of a directional contribution in applying it, the underlying structures, including muscles which determine the placement of the body parts in space, and also their relations to each other, come a little closer to the normal.

Muscle energy techniques, which involve passive and active stretching of shortened and often fibrosed structures, have marked effects on fascial changes such as those hinted at by Rolf, and which have universal involvement in total body function, as indicated by osteopathic physician Angus Cathie's list of the properties of fascia (Cathie 1974).

Fascia, he tells us:

- Is richly endowed with nerve endings
- Has the ability to contract and relax elastically
- Provides extensive muscular attachments
- Supports and stabilises all structures, so enhancing postural balance
- Is vitally involved in all aspects of movement
- Assists in circulatory economy, especially of venous and lymphatic fluids
- Will demonstrate changes preceding many chronic degenerative diseases
- Will frequently be associated with chronic passive tissue congestion when such changes occur

- Will respond to tissue congestion by formation of fibrous tissue, followed by increased hydrogen ion concentration in articular and periarticular structures
- Will form specialised 'stress bands' in response to the load demanded of it
- Commonly produces a pain of a burning nature in response to sudden stress-trauma
- Is a major arena of many inflammatory processes
- Is the medium along the fascial planes of which many fluids and infectious processes pass
- Is the tissue which surrounds the CNS.

Cathie also points out that many 'trigger' spots correspond to sites where nerves pierce fascial investments. Stress on the fascia can be seen to result from faulty muscular patterns of use, altered bony relationships, altered visceral position and postural imbalance, whether of a sustained nature or violently induced by trauma.

It is safe to say that there are no musculo-skeletal problems which do not involve fascia and, since it is a continuous structure throughout the body, any alterations in its structural integrity, by virtue of tensions, shortening, thickening or calcification, are bound to impact on areas at a distance from the site of the stress.

Fascia and posture

The specialised fascial structures – plantar, iliotibial, lumbodorsal, cervical and cranial – stabilise the body and permit an easier maintenance of the upright position, and these are among the first to show signs of change in response to postural defects.

Korr (1986) once again, as in so much of his writing, sums up what we know in a manner which enlightens further:

While biomechanical dysfunction is usually viewed as a causative or contributing factor in the patient's problem, it is itself a consequence of the imperfections in that person's total adaptation to the relentless force of gravity . . . It is no semantic accident that 'posture' and 'attitude' apply to both the physical and psychological domains. Given the unity of the body and mind, posture reflects the history and status of both and helps in determining where and how the body framework is vulnerable.

What causes abnormal fascial tension?

Cisler (1994) summarises the commonest factors which produce fascial stress as:

- Faulty muscular activity
- Altered position of fascia in response to osseous changes
- Changes in visceral position (ptosis)
- Sudden or gradual alterations in vertebral mechanics.

He also tells us that, 'In specific regions, where fascial tension is great due to associated muscular attachments, or closely related articulations, skeletal disorders are likely to be the site of a marked, burning type of pain in localised fascia.' Changes in the fascia can result from passive congestion which results in fibrous infiltration and a more 'sol'-like consistency than is the norm. Under healthy conditions a 'gel'-like ground substance follows the laws of fluid mechanics. Clearly the more resistive drag there is in a colloidal substance the greater will be the difficulty in normalising this.

Scariati (1991) points out that colloids are not rigid; they conform to the shape of their container, and respond to pressure even though they are not compressible. The amount of resistance they offer increases proportionally to the velocity of motion applied to them, which makes a gentle touch a fundamental requirement if viscous drag and resistance is to be avoided when attempting to produce a release.

When stressful forces (either undesirable or therapeutic) are applied to fascia there is a first reaction in which a degree of slack is allowed to be taken up, followed by what is colloquially referred to as 'creep' – a variable degree of resistance (depending upon the state of the tissues). Creep is an honest term which accurately describes the slow, delayed yet continuous stretch which occurs in response to a continuously applied load, as long as this is gentle enough to not provoke the resistance of colloidal 'drag'. This highlights the absolute

need in applying MET (as will be described in later chapters), for stretching to be slow and gentle, involving 'taking out of slack', followed by stretch at the pace the tissues allow, unforced, if a defensive response is to be avoided.

Since the fascia comprises a single structure, the implications for body-wide repercussions of distortions in that structure are clear. An example of one possible negative influence of this sort is to be found in the fascial divisions within the cranium, the tentorium cerebelli and falx cerebri which are commonly warped during birthing difficulties (too long or too short a time in the birth canal, forceps delivery etc.) and which are noted in craniosacral therapy as affecting total body mechanics via their influence on fascia (and therefore the musculature) throughout the body (Brookes 1984).

Cantu (1992) describes what he sees as the 'unique' feature of connective tissue as its 'deformation characteristics'. This refers to a combined viscous (permanent) deformation characteristic, as well as an elastic (temporary) deformation characteristic. This leads to the clinically important manner in which connective tissue responds to applied mechanical force by first changing in length, followed by some of this change being lost while some remains. The implications of this phenomenon can be seen in the application of stretching techniques to such tissues as well as in the way they respond to postural and other repetitive insults.

Such changes are not, however, permanent since collagen (the raw material of fascia/connective tissue) has a limited (300 to 500 day) half-life and, just as bone adapts to stresses imposed upon it, so will fascia. If therefore negative stresses (posture, use etc.) are modified for the better and/or positive 'stresses' are imposed – manipulation and/or exercise for example – dysfunctional connective tissue can usually be improved over time (Neuberger et al 1953).

Cantu and Grodin, in their evaluation of the myofascial complex, conclude that therapeutic approaches which sequence their treatment protocols to involve the superficial tissues (involving autonomic responses) as well as deeper tissues (influencing the mechanical components of the musculoskeletal system) and which also address the factor of mobility (movement), are in tune with the requirements of the body when dysfunctional.

Box 2.1   Stress factors leading to musculoskeletal dysfunction (Fig. 2.2)

- Acquired postural imbalances (Rolf 1977)
- 'Pattern of use' stress (occupational, recreational etc.)
- Inborn imbalance (short leg, short upper extremity, small hemipelvis, fascial distortion via birth injury etc.)
- The effects of hyper- or hypomobile joints, including arthritic changes
- Repetitive strain from hobby, recreation, sport etc. (overuse)
- Emotional stress factors (Barlow 1958)
- Trauma (abuse), inflammation and subsequent fibrosis
- Disuse, immobilisation
- Reflexogenic influences (viscerosomatic, myofascial and other reflex inputs) (Beal 1983)
- Climatic stress such as chilling
- Nutritional imbalances (vitamin C deficiency reduces collagen efficiency for example) (Pauling 1976)
- Infection.

THE EVOLUTION OF MUSCULOSKELETAL DYSFUNCTION
(Guyton 1987, Janda 1985, Lewit 1974)

The normal response of muscle to any form of stress is to increase in tone (Barlow 1958, Selye 1956). Some of the stress factors which negatively influence musculoskeletal soft tissues structure or function, producing irritation, increased muscle tension and pain, are listed in Box 2.1.

A chain reaction will evolve as any one, or combination of, the stress factors listed in Box 2.1, or additional stress factors, cumulatively demand increased muscular tone in those structures obliged to compensate for, or adapt to them, resulting in the following events:

- The muscles antagonistic to the hypertonic muscles become weaker (inhibited) – as may the hypertonic muscles themselves.
- The stressed muscles develop areas of relative hypoxia and ultimately ischaemia while, simultaneously, there will be a reduction in the efficiency with which metabolic wastes are removed.
- The combined effect of toxic build-up (largely the by-products of the tissues themselves)

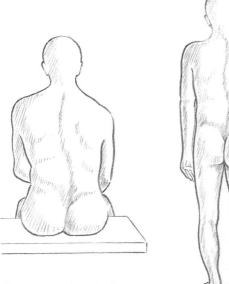
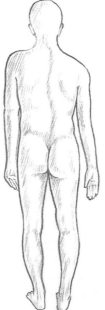
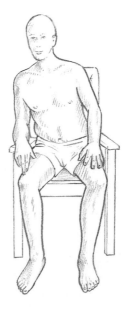

Figure 2.2   Examples of common congenital structural imbalances which result in sustained functional/postural stress – small hemipelvis, short leg and short upper extremity.

(Cyriax 1962) and oxygen deprivation leads to irritation, sensitivity and pain, which feeds back into the loop, so creating more hypertonicity and pain. This feedback loop becomes self-perpetuating.

- Oedema may also be a part of the response of the soft tissues to stress.
- If inflammation is part of the process, fibrotic changes in connective tissue may follow.
- Neural structures in the area may become facilitated, and therefore hyperreactive to stimuli, further adding to the imbalance and dysfunction of the region (see discussion later in this chapter, of myofascial trigger points and other areas of facilitation, pp 38–40).
- Initially, the soft tissues involved will show a reflex resistance to stretch and after some weeks (some say less, see Van Buskirk's view earlier in this chapter, p. 19) a degree of fibrous infiltration may appear as the tissues under greatest stress mechanically, and via oxygen lack, adapt to the situation.
- The tendons and insertions of the hypertonic muscles will also become stressed and pain and localised changes will begin to manifest in these regions. Tendon pain and periosteal discomfort are noted (Lewit & Simons 1984).
- If any of the hypertonic structures cross joints, and many do, these become crowded and some degree of imbalance will manifest, as abnormal movement patterns evolve (with antagonistic and synergistically related muscles being excessively hypertonic and/or hypotonic, for example) leading ultimately to joint dysfunction.
- Localised reflexively active structures (trigger points) will emerge in the highly stressed, most ischaemic, tissues, and these will themselves become responsible for the development of new dysfunction at distant target sites, typically inhibiting antagonist muscles (Travell & Simons 1983, Lewit & Simons 1984).
- Because of excessive hypertonic activity there will be energy wastage and a tendency to fatigue – both locally and generally (Gutstein 1955).
- Functional imbalances will occur, for example involving respiration, when chain reactions of hypertonicity and weakness impact on this vital function (Garland 1994, Lewit 1980).
- Muscles will become involved in 'chain reactions' of dysfunction. A process develops in which some muscles will be used inappropriately as they learn to compensate for other structures which are weak or restricted, leading to adaptive movements, and loss of the ability to act synergistically as in normal situations (see notes on 'crossed syndromes' below, pp 32–33, and also Dr Liebenson's comments in Chapter 5, p. 109) (Janda 1985).
- Over time, the central nervous system learns to accept altered patterns of use as normal, adding further to the complication of recovery since rehabilitation will now demand a relearning process as well as the more obvious structural (shortness) and functional (inhibition/weakness) corrections (Knott & Voss 1968).

Fitness, weakness, strength and hypermobility influences

While much of the emphasis in the rationale of use of MET relates to hypertonic structures it would be folly to neglect to mention the converse – hypotonia. In particular, Kraus (1970) presented evidence of the negative influence of relative lack of fitness on the evolution of low back pain.

Whether through acquired lack of fitness, reflex inhibition, or more seriously, inborn hypomobility, the fact is that lack of tone contributes enormously to musculoskeletal problems, imbalances and changes in functional sequence patterns, and generally causes a good deal of compensating overuse by synergistic or related muscles (Fahrni 1966, Janda 1960). Janda (1986a) describes weakness in muscles which relate to altered movement patterns, resulting from 'changed motor regulation and motor performance.' Structural and functional factors can be involved in a variety of complex ways. Janda says, 'A motor defect [weakness] of a neurological origin can almost always be considered as a result of the combination of a direct structural (morphological) lesion of some motor neurons and of inhibition effects. Both causes may occur even in the same neuron.'

Deterioration of muscle function can be demonstrated by three syndromes, according to Janda:

- Hypotonia which can be determined by inspection and palpation
- Decrease in strength which can be determined by testing, (although, according to Janda, evaluation of strength is 'difficult and inaccurate as it is often impossible to differentiate the function of individual muscles')
- Changed sequence of activation in principal movement patterns, which can be more easily observed and evaluated if they are well understood (see also Chapter 5).

Ligaments and muscles which are hypermobile do not adequately protect joints and therefore fail to prevent excessive ranges of motion from being explored. Without this stability, overuse and injury stresses evolve and muscular overuse is inevitable. Janda observes that in his experience, 'In races in which hypermobility is common there is a prevalence of muscular and tendon pain, whereas typical back pain or sciatica are rare.'

Logically, the excessive work rate of muscles which are adopting the role of 'pseudo-ligaments' leads to tendon stress and muscle dysfunction, increasing tone in the antagonists of whatever is already weakened and complicating an already complex set of imbalances, including altered patterns of movement (Beighton et al 1983, Janda 1984).

Characteristics of altered movement patterns

Among the key alterations which are demonstrable in patterns of altered muscle movement are:

- The start of a muscle's activation is delayed, resulting in an alteration in the order in which a sequence of muscles is activated.
- Noninhibited synergists or stabilisers often

activate earlier in the sequence than the inhibited, weak muscle.
- There is an overall decrease in activity in the affected muscle, which in extreme cases can result in EMG readings showing it to be almost completely silent. This can lead to a misinterpretation, that muscle strength is totally lacking when in fact, after proper facilitation, it may be capable of being activated towards more normal function. Janda calls these changes 'pseudoparesis'.
- An anomalous response is possible from such muscles since, unlike the usually beneficial activation of motor units seen in isometric training, such work against resistance can actually decrease even further the activity of pseudoparetic muscles (similar to the effect seen in muscles which are antagonists of the muscles in spasm in poliomyelitis).
- Some muscles are more likely to be affected by hypotonia, loss of strength and the effects of altered movement patterns. Janda points to tibialis anticus, peronei, vasti, long thigh adductors, the glutei, the abdominal muscles, the lower stabilisers of the scapula, the deep neck flexors.

Among the causes of such changes in mainly phasic muscles are the effects of reciprocal inhibition by tight muscles, and in such cases Janda comments, 'Stretching and achievement of normal length of the tight muscles disinhibits the pseudoparetic muscles and improves their activity.'

The phenomenon of increased tone is the other side of the picture.

What does increased bind/tone actually represent?

Janda notes that the word 'spasm' is commonly used without attention to various functional causes of hypertonicity and he has divided this phenomenon into five variants (Janda 1989):

1. Hypertonicity of limbic system origin which may be accompanied by evidence of stress, and be associated with, for example, tension-type headaches.

2. Hypertonicity of a segmental origin, involving interneuron influence. The muscle is likely to be spontaneously painful, and will probably be painful to stretch and will certainly have weak (inhibited) antagonists.

3. Hypertonicity due to uncoordinated muscle contraction resulting from myofascial trigger point activity. The muscle will be painful spontaneously if triggers are active. There may only be increased tone in part of the muscle which will be hyperirritable while neighbouring areas of the same muscle may be inhibited.

4. Hypertonicity resulting from direct pain irritation, such as might occur in torticollis. This muscle would be painful at rest, not only when palpated and would demonstrate electromyographic evidence of increased activity even at rest. This could be described as reflex spasm due to nociceptive influence (see above, p. 19, for more on nociceptive influences).

5. Overuse hypertonicity results in muscles becoming increasingly irritable, with reduced range of motion, tightness and painful only on palpation.

Thus increased tone of functional origin can result from pain sources, from trigger point activity, from higher centres or CNS influences and from overuse.

Liebenson (1990a) suggests that each type of hypertonicity requires different therapeutic approaches, ranging from adjustment (joint manipulation) through use of soft tissue and rehabilitation and facilitation approaches. The many different MET variations offer the opportunity to influence all stages of dysfunction, as listed above – the acute, the chronic and everything in between – as will become clear in our evaluation of the methods.

Lee's description of postural adaptation

Diane Lee (Grieve 1986) explains how a patient presenting with pain, loss of functional movement or altered patterns of strength, power or endurance will probably either have suffered a major trauma which has overwhelmed the physiological limits of relatively healthy tissues, or will be displaying 'gradual decompensation demonstrating slow exhaustion of the tissue's adaptive potential, with or without trauma'.

She shows how progressive postural adaptation, influenced by time factors, and possibly by trauma, leads to exhaustion of the body's adaptive potential, resulting in dysfunction and, ultimately, symptoms. She reminds us of Hooke's Law which states that, within the elastic limits of any substance, the ratio of the stress applied to the strain produced is constant (Bennet 1952).

In simple terms this means that tissue capable of deformation will absorb, or adapt to, forces applied to it within its elastic limits, beyond which it will break down or fail to compensate. Lee rightly reminds us that, while attention to specific tissues incriminated in producing symptoms often gives excellent short-term results, 'unless treatment is also focused towards restoring function in asymptomatic tissues responsible for the original postural adaptation and subsequent decompensation, the symptoms will recur.'

Clearly we all adapt and (de)compensate at our own rates, depending upon multiple variables ranging from our inherited tendencies, genetic make-up and nutritional status, to the degree, variety and intensity of the stressors confronting us, past and present.

Adding to the complexity of these responses is yet one more variable; the fact that there are predictable and palpable differences in the responses of the soft tissues to stress – some muscles becoming progressively weak while others become progressively hypertonic (Janda 1978).

DIFFERENT STRESS RESPONSE OF POSTURAL AND PHASIC MUSCLES

One of the most important revelations over the past two decades has come from research by Lewit, Korr, Janda, Basmajian, and others, which shows that muscles which have predominantly stabilising functions will shorten when stressed while others which have more active 'moving' or phasic functions will not shorten but will become weak (inhibited) (Basmajian 1978, Janda 1983, Janda 1988, Korr 1980, Lewit 1992).

The muscles which shorten are those which have a primarily postural rather than phasic (active, moving) role and it is possible to learn to conduct, in a short space of time (10 minutes or so) an assessment sequence in which the majority of these can be identified as being either short or normal (Chaitow 1991a).

Janda informs us that postural muscles have a tendency to shorten, not only under pathological conditions but often under normal circumstances. He has noted, using electromyographic instrumentation, that 85% of the walking cycle is spent on one leg or the other, and that this is the most common postural position for man. Those muscles which enable this position to be satisfactorily adopted (one-legged standing) are genetically older; they have different physiological, and probably biochemical, qualities compared with phasic muscles which normally weaken and exhibit signs of inhibition in response to stress or pathology.

Postural muscles

Those postural muscles which respond to stress by shortening are listed in Box 2.2. The scalenes are a borderline set of muscles – which start life as phasic muscles but which can become through overuse/abuse more postural in their function (Fig. 2.3).

Can postural muscles/phasic muscles change from one form into the other?

While Lewit and Janda (Lewit 1992b) have shown that postural muscles under stress will

---

Box 2.2   Postural muscles that shorten under stress

Gastrocnemius, soleus, medial hamstrings, short adductors of the thigh, hamstrings, psoas, piriformis, tensor fascia lata, quadratus lumborum, erector spinae muscles, latissimus dorsi, upper trapezius, sternomastoid, levator scapulae, pectoralis major and the flexors of the arms.

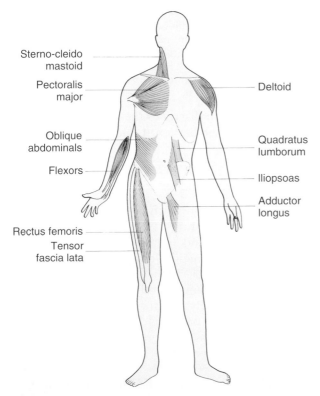

Figure 2.3A   The major postural muscles of the anterior aspect of the body.

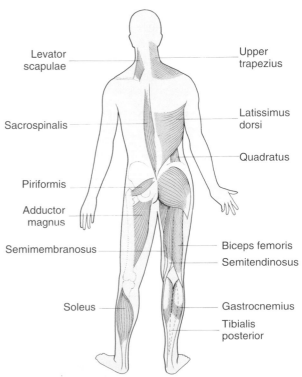

Figure 2.3B   The major postural muscles of the posterior aspect of the body.

shorten and phasic muscles similarly stressed will weaken, it is now becoming clear that the function of a muscle can be modified, which helps to explain some mysteries – for example, why the scalenes are sometimes short and sometimes weak and sometimes both, and yet are classified generally as phasic muscles and sometimes as 'equivocal' (maybe postural and maybe phasic).

Lin (Lin et al 1994), writing in *The Lancet*, examined motor muscle physiology in growing children, reviewing current understanding of the postural/phasic muscle interaction:

Muscles are considered to be developmentally static, which is surprising considering in-vitro information on the development and adaptability of muscles derived from mammals. Buller [1960] and Eccles showed that a committed muscle-fibre type could be transformed from slow-twitch to fast-twitch and vice-versa in their cross innervation experiments, confirming that impulse traffic down the nerve conditions the fibre type.

The implication of this is that if a group of muscles such as the scalenes are dedicated to movement – which they should be – and not to stabilisation, which they may have to be if postural stresses are imposed, they can become 'postural' in type, and so will develop a tendency to shorten if stressed.

Characteristics of postural and phasic muscles

The characteristics which identify a muscle as belonging to one or other of these two groups are given in Table 2.1.

All muscles comprise both red and white, fast and slow fibres which produce both postural and phasic functions; however, the classification

Table 2.1 Postural/phasic muscle characteristics

	Postural muscles	Phasic muscles
Type	Slow twitch – red	Fast twitch – white
Respiration	Anaerobic	Aerobic
Function	Static/supportive	Phasic/active
Dysfunction	Shorten	Weaken
Treatment	Stretch/relax	Facilitate/strengthen

of a muscle into either a 'postural' or 'phasic' group is made on the basis of their predominant activity, their major functional tendency.

Rehabilitation implications

Janda has shown that before any attempt is made to strengthen weak muscles, any hypertonicity in their antagonists should be addressed by appropriate treatment which relaxes them – for example, by stretching using MET. Relaxation of hypertonic muscles leads to an automatic regaining of strength of their antagonists once inhibitory hypertonic effects have been removed. Should the hypertonic muscle also be weak, it commonly regains strength following stretch/relaxation (Janda 1978). Commenting on this phenomenon, chiropractic rehabilitation expert Craig Liebenson (1990b) states:

Once joint movement is free, hypertonic muscles relaxed, and connective tissue lengthened, a muscle-strengthening and movement coordination program can begin. It is important not to commence strengthening too soon because tight, overactive muscles reflexively inhibit their antagonists, thereby altering basic movement patterns. It is inappropriate to initiate muscle strengthening programs while movement performance is disturbed, since the patient will achieve strength gains by use of 'trick' movements.

(Dr Liebenson discusses these and other treatment and rehabilitation topics more fully in Chapter 5.)

A common phenomenon

Just how common such imbalances are was illustrated by Schmid (1984), who studied the main postural and phasic muscles in eight members of the male Olympic ski teams from Switzerland and Liechtenstein.

He found that amongst this group of superbly fit individuals fully 6 of the 8 had demonstrably short right iliopsoas muscles, while 5 of the 8 also had left iliopsoas shortness and the majority also displayed weakness of the rectus abdominus muscles. A number of other muscle imbalances were noted and the conclusion was that athletic fitness offers no more protection from muscular dysfunction than does a sedentary lifestyle.

Liebenson (1990b) discusses the work of Sommer (1985) who found that competitive basketball and volleyball players frequently produce patellar tendonitis and other forms of knee dysfunction due to the particular stresses they endure because of muscular imbalances. Their ability to jump is often seriously impaired by virtue of shortened psoas and quadriceps muscles with associated weakness of gluteus maximus. This imbalance leads to decreased hip extension and hyperextension of the knee joint. Once muscular balance is restored, a more controlled jump is possible as is a reduction in reported fatigue.

The element of fatigue should not be forgotten in this equation, since hypertonic muscles are working excessively both to perform their functions and often to compensate for weakness in associated muscles.

Kuchera and associates (Kuchera et al 1990) have shown that in healthy collegiate volunteers a significant correlation exists between a history of trauma and the type of athletic activity pursued, most notably in the golf team who displayed a rotation to the right around the right oblique sacral axis. The volunteers were subjected to a variety of assessments including palpatory structural analysis, anthropomorphic measurements, radiographic series as well as photographic centre of gravity analyses. Well-compensated patterns of fascia were noted in those who had a low incidence of back pain, whereas, conversely, a higher incidence of non-compensated patterning related to back pain within the previous year. Subjects reporting a significant history of psoas muscle problems were found to have a high incidence of non-compensated fascial patterning.

When is a condition most suitable for MET application?

Evjenth (Evjenth and Hamberg 1984) succinctly summarises: 'Every patient with symptoms involving the locomotor system, particularly symptoms of pain and/or constrained move-ment, should be examined to assess joint and muscle function. If examination shows joint play to be normal, but reveals shortened muscles or muscle spasm, then treatment by stretching [and by implication MET] is indicated.'

Where do joints fit into the picture?

Janda has an answer to the emotive question when he says that it is not known whether dysfunction of muscles causes joint dysfunction or vice versa (Janda 1988). He points out, how-ever, that since clinical evidence abounds that joint mobilisation (thrust or gentle mobilisation) influences the muscles which are in anatomic or functional relationships with the joint, it may well be that normalisation of the excessive tone of the muscles in this way is what is providing the benefit, and that, by implication, normalis-ation of the muscle tone by other means (such as MET) would provide an equally useful basis for a beneficial outcome and joint normalisation. Since reduction in muscle spasm/contraction commonly results in a reduction in joint pain, the answer to many such problems would seem to lie in appropriate soft tissue attention.

Liebenson (1990b) takes a view with a chiro-practic bias: 'The chief abnormalities of (musculo-skeletal) function include muscular hypertonicity and joint blockage. Since these abnormalities are functional rather than structural they are reversible in nature ... Once a particular joint has lost its normal range of motion, the muscles around that joint will attempt to minimise stress at the involved segment.'

After describing the processes of progressive compensation as some muscles become hyper-tonic while inhibiting their antagonists, he con-tinues, 'What may begin as a simple restriction of movement in a joint can lead to the dev-elopment of muscular imbalances and postural change. This chain of events is an example of what we try to prevent through adjustments of subluxations.'

We are left then with one view which has it that muscle release will frequently normalise joint restrictions, as well as a view which holds the opposite – that joint normalisation sorts out soft tissue problems, leaving direct work on muscles for rehabilitation settings and for attention if joint mobilisation fails to deal with long-term changes (fibrosis etc.).

It is possible that both views are to some extent correct. However, what emphasis each therapist/ practitioner gives to their prime focus – be it joint or be it soft tissues – the certainty is that what is required is anything but a purely local view, as Janda helps us to understand.

PATTERNS OF DYSFUNCTION

When a chain reaction evolves in which some muscles shorten and others weaken, predictable patterns involving imbalances develop, and Czech researcher Vladimir Janda describes the so-called upper and lower 'crossed' syndromes (see below and Tables 2.2 and 2.3).

Upper crossed syndrome (Fig. 2.4)

This involves the basic imbalance shown in Table 2.2.

Table 2.2 Upper crossed syndrome

Pectoralis major and minor	All tighten and shorten
Upper trapezius	
Levator Scapulae	
Sternomastoid	
while	
Lower and Middle trapezius	All weaken
Serratus anterior and Rhomboids	

As the changes listed in Table 2.2 take place, they alter the relative positions of the head, neck and shoulders as follows:

1. The occiput and C1/2 will hyperextend, with the head being pushed forward.
2. The lower cervical to 4th thoracic vertebrae will be posturally stressed as a result.
3. Rotation and abduction of the scapulae occurs.
4. An altered direction of the axis of the glenoid fossa will develop resulting in the humerus needing to be stabilised by additional levator scapula and upper trapezius activity, with additional activity from supraspinatus as well.

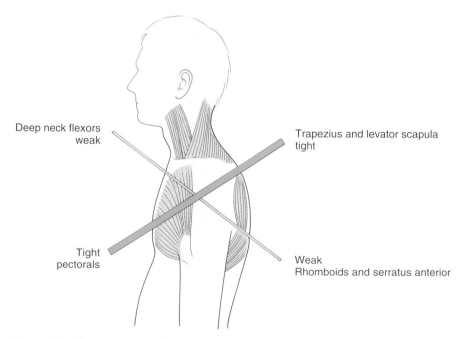

Deep neck flexors weak

Trapezius and levator scapula tight

Tight pectorals

Weak Rhomboids and serratus anterior

Figure 2.4 The upper crossed syndrome, as described by Janda.

The result of these changes is greater cervical segment strain plus referred pain to the chest, shoulders and arms. Pain mimicking angina may be noted plus a decline in respiratory efficiency.

The solution, according to Janda, is to be able to identify the shortened structures and to release (stretch and relax) them, followed by re-education towards more appropriate function.

Lower crossed syndrome (Fig. 2.5)

This involves the basic imbalance shown in Table 2.3.

The result of the chain reaction in Table 2.3 is that the pelvis tips forward on the frontal plane, flexing the hip joints and producing lumbar lordosis and stress at L5–S1 with pain and irritation. A further stress commonly appears in the sagittal plane in which quadratus lumborum tightens and gluteus maximus and medius weaken.

When this 'lateral corset' becomes unstable, the pelvis is held in increased elevation, accentuated when walking, resulting in L5–S1 stress in the sagittal plane. One result is low back pain. The combined stresses described produce instability at the lumbodorsal junction, an unstable transition point at best.

Also commonly involved are the piriformis muscles which in 20% of individuals are penetrated by the sciatic nerve so that piriformis syndrome can produce direct sciatic pressure and pain. Arterial involvement of piriformis shortness produces ischaemia of the lower extremity, and through a relative fixation of the sacrum, sacroiliac dysfunction and pain in the hip.

The solution for an all too common pattern such as this is to identify the shortened structures and to release them, ideally using variations on the theme of MET, followed by re-education of posture and use.

Chain reaction leads to facial and jaw pain

In case it is thought that such imbalances are of merely academic interest, a practical example of the negative effects of the chain reactions described above is given by Janda (1986b) in an article entitled 'Some aspects of extracranial causes of facial pain'.

Janda's premise is that temporomandibular joint (TMJ) problems and facial pain can be analysed in relation to the patient's whole posture. He has hypothesised that the muscular pattern associated with TMJ problems may be considered as locally involving hyperactivity and tension in the temporal and masseter muscles while, because of this hypertonicity, reciprocal inhibition occurs in the suprahyoid, digastric and mylohyoid muscles. The external pterygoid in particular often develops spasm. This imbalance between jaw adductors and jaw openers alters the ideal position of the condyle and leads to a consequent redistribution of stress on the joint, leading to degenerative changes.

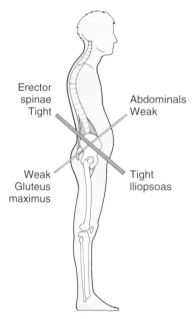

Erector spinae Tight

Abdominals Weak

Weak Gluteus maximus

Tight Iliopsoas

Figure 2.5 The lower crossed syndrome, as described by Janda.

Table 2.3 Lower crossed syndrome

Hip flexors	All tighten and shorten
Iliopsoas, Rectus femoris	
TFL, Short adductors	
Erector spinae group of the trunk	
while	
Abdominal and gluteal muscles	All weaken

Janda describes the typical pattern of muscular dysfunction of an individual with TMJ problems as involving upper trapezius, levator scapula, scaleni, sternomastoid, suprahyoid, lateral and medial pterygoid, masseter and temporal muscles, all of which show a tendency to tighten and to develop spasm.

He notes that while the scalenes are unpredictable, and while commonly, under overload conditions, they become atrophied and weak, they may also develop spasm, tenderness and trigger points.

The postural pattern in a TMJ patient might involve (Fig. 2.6; see also Fig. 2.5):

1. Hyperextension of knee joints
2. Increased anterior tilt of pelvis
3. Pronounced flexion of hip joints
4. Hyperlordosis of lumbar spine
5. Rounded shoulders and winged (rotated and abducted) scapulae
6. Cervical hyperlordosis
7. Forward thrust of head
8. Compensatory over-activity of upper trapezius and levator scapulae
9. Forward thrust of head resulting in opening of mouth and retraction of mandible.

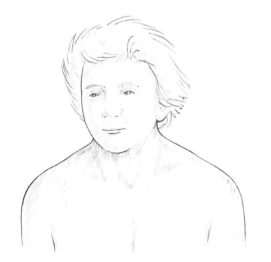

Figure 2.6 A typical pattern of upper thoracic and cervical stress as described by Janda would involve a degree of TMJ stress. Note the 'gothic shoulders' which result from upper trapezius hypertonicity and shortening.

This series of changes provokes increased activity of the jaw adductor and protractor muscles, creating a vicious cycle of dysfunctional activity. Intervertebral joint stress in the cervical spine follows.

The message which can be derived from this evidence is that patterns first need to be identified before they can be assessed for the role they might be playing in the patient's pain and restriction conditions, and certainly before these can be successfully and appropriately treated. (Dr Liebenson discusses a number of Vladimir Janda's observation assessment methods in Chapter 5.)

Patterns of change with inappropriate breathing (Fig. 2.7)

Garland (1994) describes the somatic changes which follow from a pattern of hyperventilation, upper chest breathing.

When faced with persistent upper chest breathing patterns we should be able to identify reduced diaphragmatic efficiency and commensurate restriction of the lower rib cage as these evolve into a series of changes with accessory breathing muscles being inappropriately and excessively used:

- A degree of visceral stasis and pelvic floor weakness will develop, as will an imbalance between increasingly weak abdominal muscles and increasingly tight erector spinae muscles.
- Fascial restriction from the central tendon via the pericardial fascia, all the way up to the basi-occiput, will be noted.
- The upper ribs will be elevated and there will be sensitive costal cartilage tension.
- The thoracic spine will be disturbed by virtue of the lack of normal motion of the articulation with the ribs, and sympathetic outflow from this area may be affected.
- Accessory muscle hypertonia, notably affecting the scalenes, upper trapezius and levator scapulae will be palpable and observable.
- Fibrosis will develop in these muscles as will myofascial trigger points.
- The cervical spine will become progressively

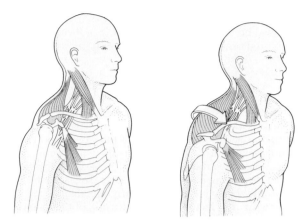

Figure 2.7A   A progressive pattern of postural and biomechanical dysfunction develops resulting in, and aggravated by, inappropriate breathing function.

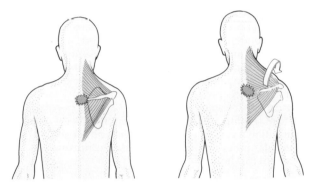

Figure 2.7B   The local changes in the muscles of an area being stressed in this way will include the evolution of fibrotic changes and myofascial trigger points.

rigid with a fixed lordosis being a common feature in the lower cervical spine.

- A reduction in the mobility of the 2nd cervical segment and disturbance of vagal outflow from this region is likely.
- Although not noted in Garland's list of dysfunction (in which he states, 'psychology overwhelms physiology'), we should bear in mind that the other changes which Janda has listed in his upper crossed syndrome (p. 32) are also likely consequences, including the potentially devastating effects on shoulder function of the altered position of the scapulae and glenoid fossae as this pattern evolves.
- Also worth noting in relation to breathing function and dysfunction are two important

muscles not included in Garland's list; quadratus lumborum and iliopsoas, both of which merge fibres with the diaphragm. Since these are both postural muscles, with a propensity to shortening when stressed, the impact of such shortening, either uni- or bilaterally, can be seen to have major implications for respiratory function, whether the primary feature of such a dysfunction lies in diaphragmatic or in muscular distress.

Recall that amongst possible stress factors which will result in shortening of postural muscles is disuse, and that a situation in which upper chest breathing has replaced diaphragmatic breathing as the norm would lead to reduced diaphragmatic excursion and consequent reduction in activity for those aspects of quadratus lumborum and psoas which are integral with it, and shortening would be a likely result of this.

Recall also Schmid's skiers, whose most obvious shortening feature was psoas. It is hard to imagine that their overall performance efficiency would not be impaired by the impact of bilaterally short psoas structures (as found in 75% of those examined) dragging on their diaphragms as they hurtle down mountainsides.

Garland concludes his listing of somatic changes associated with hyperventilation by saying, 'Physically and physiologically [all of] this runs against a biologically sustainable pattern, and in a vicious cycle, abnormal function (use) alters normal structure, which disallows return to normal function.' He also points to the likelihood of counselling (for associated anxiety or depression, perhaps) and breathing retraining, being far more likely to be successfully initiated if the structural component(s) – as listed – are dealt with in such a way as to minimise the effects of the somatic changes described.

Pioneer osteopathic physician Carl McConnell (1902) reminds us of wider implications:

Remember that the functional status of the diaphragm is probably the most powerful mechanism of the whole body. It not only mechanically engages the tissues of the pharynx to the perineum, several times per minute, but is physiologically indispensable to the activity of every cell in the body. A working knowledge of the crura, tendon, and the extensive ramification of the diaphragmatic tissues, graphically depicts the significance of structural continuity and functional unity. The wealth of soft tissue work centring in the powerful mechanism is beyond compute, and clinically it is very practical.

Fascia and the thorax

In both Garland's and McConnell's discussion of respiratory function mention has been made of fascia, the importance of which was indicated earlier in this chapter when some of Rolf's and Cathie's thoughts were enumerated. An additional reference to the ubiquitous nature and vital importance of this structure comes from Leon Page (1952), who discusses the involvement of fascia in the thoracic region:

The cervical fascia extends from the base of the skull to the mediastinum and forms compartments enclosing oesophagus, trachea, carotid vessels and provides support for the pharynx, larynx and thyroid gland. There is direct continuity of fascia from the apex of the diaphragm to the base of the skull, extending through the fibrous pericardium upward through the deep cervical fascia and the continuity extends not only to the outer surface of the sphenoid, occipital and temporal bones but proceeds further through the foramina in the base of the skull around the vessels and nerves to join the dura.

Goldthwaite's overview

Goldthwaite, in his classic 1930s discussion of posture, links a wide array of problems to the absence of balanced posture (Goldthwaite 1945). Clearly some of what he hypothesises remains conjecture, but we can see just how much impact postural stress can have on associated tissues, starting with diaphragmatic weakness:

The main factors which determine the maintenance of the abdominal viscera in position are the diaphragm and the abdominal muscles, both of which are relaxed and cease to support in faulty posture. The disturbances of circulation resulting from a low diaphragm and ptosis may give rise to chronic passive congestion in one or all of the organs of the abdomen and pelvis, since the local as well as general venous drainage may be impeded by the failure of the diaphragmatic pump to do its full work in the drooped body. Furthermore, the drag of these congested organs on their nerve supply, as well as the pressure on the sympathetic ganglia and plexuses, probably causes many irregularities in their function, varying from partial paralysis to overstimulation. All these organs receive fibres from both the vagus and sympathetic systems, either one of which may be disturbed. It is probable that one or all of these factors are active at various times in both the stocky and the slender anatomic types, and are responsible for many functional digestive disturbances. These disturbances, if continued long enough may lead to diseases later in life. Faulty body mechanics in early life, then, becomes a vital factor in the production of the vicious cycle of chronic diseases and presents a chief point of attack in its prevention . . . In this upright position, as one becomes older, the tendency is for the abdomen to relax and sag more and more, allowing a ptosic condition of the abdominal and pelvic organs unless the supporting lower abdominal muscles are taught to contract properly. As the abdomen relaxes, there is a great tendency towards a drooped chest, with narrow rib angle, forward shoulders, prominent shoulder blades, a forward position of the head, and probably pronated feet. When the human machine is out of balance, physiological function cannot be perfect; muscles and ligaments are in an abnormal state of tension and strain. A well-poised body means a machine working perfectly, with the least amount of muscular effort, and therefore better health and strength for daily life.

Note how closely Goldthwaite mirrors the picture Janda paints in his upper and lower crossed syndromes (pp 32–33), and 'posture and facial pain' description (p. 33).

Patterns of function and patterns of dysfunction are seen in the various examples given to provide us with fertile soil in which to seek a crop of dysfunctional tissues, ripe for therapeutic harvesting.

Korr's trophic influence research

Irwin Korr has spent half a century investigating the scientific background to osteopathic methodology and theory, and amongst his most important research has been that which demonstrated the role of neural structures in delivery of trophic substances (Korr 1986, Korr et al 1967). The various patterns of stress which we have covered in this chapter are capable of drastically affecting this. He states:

Also involved in somatic dysfunction are neural influences that are based on the transfer of specific proteins synthesised by the neuron to the innervated tissue. This delivery is accomplished by axonal transport and junctional traversal. These 'trophic' proteins are thought to exert long-term influences on the developmental, morphologic, metabolic and functional qualities of the tissues – even on their viability. Biomechanical abnormalities in the musculoskeletal system can cause trophic disturbances in at least two ways (1) by mechanical deformation (compression, stretching, angulation, torsion) of the nerves, which impedes axonal transport; and by (2) sustained hyperactivity of neurons in facilitated segments of the spinal cord [see below, p. 38] which slows axonal transport and which because of metabolic changes, may affect protein synthesis by the neurons. It appears that manipulative treatment would alleviate such impairments of neurotrophic function.

IDENTIFICATION AND NORMALISATION OF PATTERNS OF DYSFUNCTION

Observation, palpation, specific tests – these are the ways in which such patterns may be identified and assessed so that treatment can take account of more than the local dysfunction and can place the patient's symptoms within the context of whole-body dysfunctional patterns which represent the sum of their present adaptation and compensation efforts.

Patterns of imbalance can be observed in predictable areas, relating to specific forms of dysfunction (headache, thoracic inlet, low back etc.) and the reader is directed to Dr. Liebenson's analysis of this approach to assessment in Chapter 5.

If an imbalance pattern is recognisable, and, within that, emphasis is given to what is hypertonic and what (within both hypertonic and hypotonic muscles) is reflexively active as in the case of myofascial trigger points, a therapeutic starting point is possible which leads physiologically towards the normalisation and resolution – if only partially – of the somatic dysfunction patterns currently on display.

As whatever is tense and tight to an undesirable degree is released and stretched, so will antagonists regain tone, and a degree of balance be restored. As local myofascial trigger areas are resolved, so will reflexively initiated pain and sympathetic overactivity be minimised. The stress burden will be lightened, energy will be saved, function will improve, joint stress will be reduced, exacerbation of patterns of dysfunction will be modified.

This is not the end of the story, however, since re-education as to more appropriate use is clearly the ideal long-term objective if the causes of dysfunction related to misuse, abuse or overuse of the musculoskeletal system are to be addressed. However, it is suggested that such re-education – whether postural or functional (as in breathing retraining), or sensory-motor rehabilitation where faulty motor patterns are well established – will be more successfully achieved if chronic mechanical restrictions have been minimised.

Liebenson (1990b) comments, 'In rehabilitation it is important to identify and correct overactive or shortened musculature prior to attempting a muscle strengthening regimen ... The effectiveness of any rehabilitation program is enhanced if hypertonic muscles are relaxed, and if necessary stretched, prior to initiating a strengthening program' (see Ch. 5 for an introduction to Dr Liebenson's rehabilitation methods).

There are certainly other ways of normalising hypertonicity than use of MET, even if only temporarily, such as use of inhibitory ischaemic compression (Chaitow 1991b), positional release methods (Jones 1982) or joint manipulation (Lewit 1992). Indeed, many experts in manual medicine hold that manipulation of associated joints will automatically and spontaneously resolve soft tissue hypertonicity (Janda 1978, Mennell 1952); however, neither manipulation of joints nor use of methods which do not in some way stretch the tissues will reduce and encourage towards normal those structures which have become fibrotically altered – whereas MET will do so if used appropriately.

The major researchers into myofascial trigger points, Travell and Simons, influenced by the work of Karel Lewit, also suggest that use of MET is an ideal means of normalising these centres of neurological mayhem (Travell & Simons 1992).

Trigger points

It is necessary to include a brief overview of myofascial trigger points in any consideration of patterns of dysfunction.

The reflex patterns – and facilitation

In the body, when an area is stressed repetitively and chronically, the local nerve structures in that area tend to become overexcitable, more easily activated, hyperirritable – a process known as facilitation.

There are two forms of facilitation and if we are to make sense of muscle dysfunction, we have to understand these.

Segmental facilitation
(Korr 1976, Patterson 1976)

Organ dysfunction will result in facilitation of the paraspinal structures at the level of the nerve supply to that organ. If there is any form of cardiac disease, for example, there will be a 'feedback' of impulses along these same nerves towards the spine, and the muscles alongside the spine at that upper thoracic level will become hypertonic. If the cardiac problem continues the area will become facilitated, with the nerves of the area, including those passing to the heart, becoming hyperirritable (Fig. 2.8).

Electromyographic readings of the muscles alongside the spine at the upper thoracic level would show this region to be very active compared with that above and below it, and the muscles alongside the spine at that level would be hypertonic and probably painful to pressure.

Once facilitated, if there were any additional stress impacting the individual, of any sort, whether emotional physical, chemical, climatic, mechanical or whatever, or absolutely anything which imposed stress on the person as a whole – not just this particular part of their body – there would be a marked increase in neural activity in the facilitated area and not in the rest of the spinal structures.

Korr has called such an area a 'neurological lens' – it concentrates the nerve activity to the

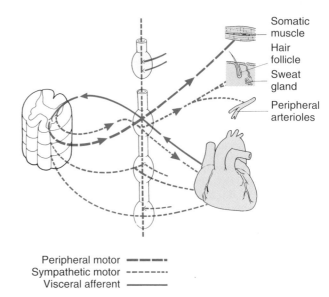

Somatic
muscle
Hair
follicle
Sweat
gland
Peripheral
arterioles

Peripheral motor
Sympathetic motor
Visceral afferent

Figure 2.8 Schematic representation of the neurological influences involved in the process of facilitation resulting from visceral dysfunction (cardiac disease in this example). Hyperirritable neural feedback to the CNS will result, which influences muscle, skin (both palpable) and venous structures in associated areas, as well as the neural supply to the organ itself.

facilitated area, so creating more activity and also a local increase in muscle tone at that level of the spine. Similar segmental (spinal) facilitation occurs in response to any organ problem, obviously affecting only the part of the spine from which the nerves to that organ emerge. Other causes of segmental (spinal) facilitation can include stress imposed on a part of the spine through injury, overactivity, repetitive stress, poor posture or structural imbalance (short leg for example).

Korr (1978) tells us that when subjects who have had facilitated segments identified 'were exposed to physical, environmental and psychological stimuli similar to those encountered in daily life, the sympathetic responses in those segments was exaggerated and prolonged. The disturbed segments behaved as though they were continually in or bordering on a state of "physiologic alarm".'

In assessing and treating somatic dysfunction,

the phenomenon of segmental facilitation needs to be borne in mind since the causes and treatment of these frequently lie outside the scope of practice of manual practitioners and therapists. In many instances, however, appropriate manipulative treatment, including use of MET, can help to 'destress' facilitated areas.

How to recognise a facilitated area

A number of observable and palpable signs indicate an area of segmental (spinal) facilitation.

Beal tells us that such an area will usually involve two or more segments – unless traumatically induced, in which case single segments are possible. The paraspinal tissues will palpate as rigid or 'board-like'. With the patient supine and the palpating hands under the patient's paraspinal area to be tested (standing at the head of the table, for example, and reaching under the shoulders for the upper thoracic area) any ceilingward 'springing' attempt on these tissues will result in a distinct lack of elasticity, unlike more normal tissues above or below the facilitated area (Beal 1983).

Grieve, Gunn and Milbrandt, and Korr have all helped to define the palpable and visual signs which accompany facilitated dysfunction (Grieve 1986, Gunn & Milbrandt 1978, Korr 1948):

- A gooseflesh appearance is observable in facilitated areas when the skin is exposed to cool air – the result of a facilitated pilomotor response.
- A palpable sense of 'drag' is noticeable as a light touch contact is made across such areas, due to increased sweat production resulting from facilitation of the sudomotor reflexes.
- There is likely to be cutaneous hyperaesthesia in the related dermatome, as the sensitivity is increased – for example, to a pin prick – due to facilitation.
- An 'orange peel' appearance is noticeable in the subcutaneous tissues when the skin is rolled over the affected segment, due to subcutaneous trophedema.
- There is commonly localised spasm of the

muscles in a facilitated area, which is palpable segmentally as well as peripherally in the related myotome. This is likely to be accompanied by an enhanced myotatic reflex due to the process of facilitation.

Local (trigger point) facilitation in muscles (Fig. 2.9)

A similar process of facilitation occurs when particularly easily stressed parts of muscle (origins and insertions for example) are overused, abused, misused, disused in any of the many ways discussed earlier in this chapter.

Localised areas of hypertonicity will develop, sometimes accompanied by some oedema, sometimes with a stringy feel – but always with a sensitivity to pressure.

Many of these palpably painful, tender, sensitive, localised, facilitated points are myofascial trigger points which are not only painful themselves when pressed, but when active will also transmit or activate pain (and other) sensations some distance away from themselves, in 'target' tissues.

In the same manner as the facilitated areas alongside the spine, these trigger points will be made more active by any stress, of whatever type, impacting on the body as a whole – not just on the area in which they lie. When not actively sending pain to a distant area, trigger

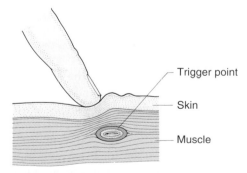

Figure 2.9 Trigger points are areas of local facilitation which can be housed in any soft tissue structure, most usually muscle and/or fascia. Palpation from the skin or at depth may be required to localise these.

points (local tender or pain areas) are said to be 'latent'. The same signs as described for spinal/ segmental facilitation can be observed and palpated in these areas, with 'drag' palpation being among the most rapid means of identifying such local dysfunction.

The leading researchers into pain, Melzack and Wall (1988), have stated that there are few, if any, chronic pain problems which do not have trigger point activity as a major part of the picture, perhaps not always as a prime cause but almost always as a maintaining feature.

What causes the trigger point to develop?

Janet Travell and David Simons are the two physicians who, above all others, have helped our understanding of trigger points. Simons has described the evolution of trigger points as follows (Lewit & Simons 1984):

In the core of the trigger lies a muscle spindle which is in trouble for some reason. Visualise a spindle like a strand of yarn in a knitted sweater . . . a metabolic crisis takes place which increases the temperature locally in the trigger point, shortens a minute part of the muscle (sarcomere) – like a snag in a sweater, and reduces the supply of oxygen and nutrients into the trigger point. During this disturbed episode an influx of calcium occurs and the muscle spindle does not have enough energy to pump the calcium outside the cell where it belongs. Thus a vicious cycle is maintained and the muscle spindle can't seem to loosen up and the affected muscle can't relax.

Simons has tested his concept and found that, at the core of trigger points, there is an oxygen deficit compared with the muscle tissue which surrounds it.

Travell has confirmed that the following factors can all help to maintain and enhance trigger point activity: nutritional deficiency (especially vitamin C, B-complex vitamins and iron); hormonal imbalances (low thyroid, menopausal or premenstrual situations for example); infections (bacteria, viruses or yeast); allergies (wheat and dairy in particular); low oxygenation of tissues (aggravated by tension, stress, inactivity, poor respiration) (Travell & Simons 1983).

Facilitation and the central nervous system

Facilitation, both segmental and local, is a feature of the shortening of muscles, as a whole, or in part.

Korr has shown that muscle spindles in areas of dysfunction are hypersensitive to change in muscle length, possibly due to incorrect spinal cord setting of the gamma-neuron control of the intrafusal muscle fibres. If influences from higher centres further exaggerate this high 'gamma-gain', exacerbation of local restriction is likely. It is only when gamma-gain is restored to normal, possibly via manipulation, that normality is achieved. This scenario may be accompanied by another feature of internal discord in which, because of multiple stresses being imposed on them, musculoskeletal reporting stations (proprioceptors in soft tissues, including skin) are presenting 'garbled' and conflicting information, so making appropriate adaptive responses impossible (Korr 1975).

An example is a situation in which high-gain spindles would be reporting greater than real muscle lengths, and this information was simultaneously being contradicted by reports from joint receptors. The CNS responses to mixed signals would be inappropriate and could lead to increased dysfunction, spasm etc. This set of events forms the basis of functional and 'strain/counterstrain' methods of soft tissue normalisation (Jones 1981).

Fibromyalgia and trigger points

Is the result of trigger point activity, known as Myofascial Pain Syndrome (MPS), the same as FibroMyalgia Syndrome (FMS)?

According to a leading researcher into this realm, P Baldry (1993), the two conditions are similar or identical in that both fibromyalgia and myofascial pain syndrome:

- Are affected by cold weather
- May involve increased sympathetic nerve activity and may involve conditions such as Raynaud's phenomenon

- Have tension headaches and paraesthesia as major associated symptoms
- Are unaffected by anti-inflammatory pain-killing medication whether of the cortisone type or standard formulations.

However, fibromyalgia and myofascial pain syndrome are different in that:

- MPS affects males and females equally, fibromyalgia mainly females.
- MPS is usually local to an area such as the neck and shoulders, or low back and legs, although it can affect a number of parts of the body at the same time – fibromyalgia is a generalised problem – often involving all four 'corners' of the body at the same time.
- Muscles which contain areas which feel 'like a tight rubber band' are found in the muscles of around 30% of people with MPS but more than 60% of people with FMS.
- People with FMS have poorer muscle endurance (they get tired faster) than do the muscles of people with MPS.
- MPS can sometimes be bad enough to cause disturbed sleep; in fibromyalgia the sleep disturbance has a more causative role, and is a pronounced feature of the condition.
- MPS produces no morning stiffness whereas fibromyalgia does.
- There is not usually fatigue associated with MPS while it is common in fibromyalgia.
- MPS can sometimes lead to depression (reactive) and anxiety whereas in a small percentage of fibromyalgia cases (some leading researchers believe) these conditions can be the trigger for the start of the condition.
- Conditions such as irritable bowel syndrome, dysmenorrhoea and a feeling of swollen joints are noted in fibromyalgia but seldom in MPS.
- Low dosage tricyclic antidepressant drugs are helpful in dealing with the sleep problems, and many of the symptoms, of fibromyalgia – but not of MPS.
- Exercise programmes (cardiovascular fitness) can help some fibromyalgia patients, according to experts, but this is not a useful approach in MPS.

- The outlook for people with MPS is excellent, since the trigger points usually respond quickly to manipulative (stretching in particular) techniques or acupuncture, whereas the outlook for fibromyalgia is less positive – with a lengthy treatment and recovery phase being the norm.

Summary
(Block 1993, Goldenberg 1993, Rothschild 1991)

Trigger points are certainly part – in some cases the major part – of the pain suffered by people with muscle pain in general as well as fibromyalgia, and when they are, MET offers a useful means of treatment, since a trigger point will reactivate if the muscle in which it lies cannot easily reach its normal resting length (Fig. 2.10).

Summary of trigger point characteristics

- Janet Travell defines trigger points as 'hyper-irritable foci lying within taut bands of muscle which are painful on compression and which refer pain or other symptoms at a distant site'.
- Embryonic trigger points will develop as satellites of existing triggers in the target area, and in time these will produce their own satellites.
- According to Professor Melzack, nearly 80% of trigger points are in exactly the same positions as known acupuncture points, as used in traditional Chinese medicine.
- Painful points which do not refer symptoms to a distant site are simply latent triggers requiring additional stress to create greater facilitation and turn them into active triggers.
- The taut band in which triggers lie will twitch if a finger is run across it, and is tight but not fibrosed, since it will soften and relax if the appropriate treatment is applied – something fibrotic tissue cannot do.
- Muscles which contain trigger points will often hurt when they are contracted (i.e. when they are working).

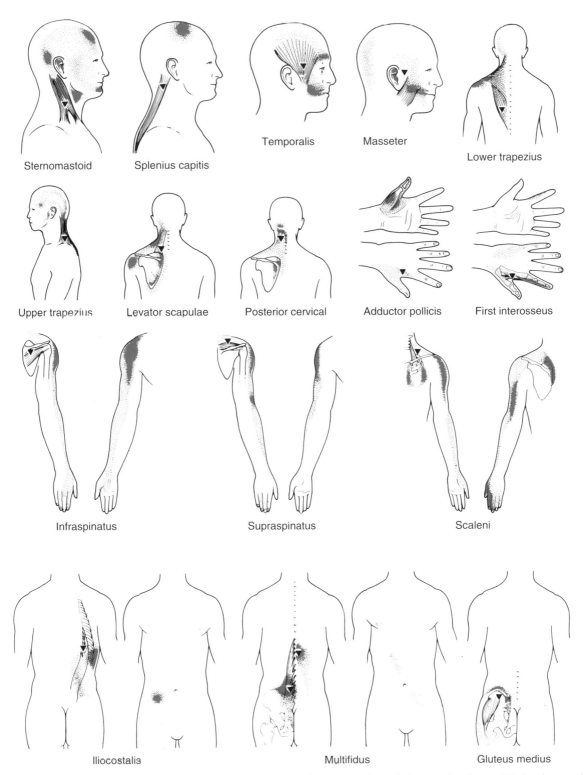

Sternomastoid

Splenius capitis

Temporalis

Masseter

Lower trapezius

Upper trapezius

Levator scapulae

Posterior cervical

Adductor pollicis

First interosseus

Infraspinatus

Supraspinatus

Scaleni

Iliocostalis

Multifidus

Gluteus medius

Figure 2.10 A selection of the most commonly found examples of representations of trigger point sites and their reference (or target) areas. Trigger points found in the same sites in different people will usually refer to the same target areas.

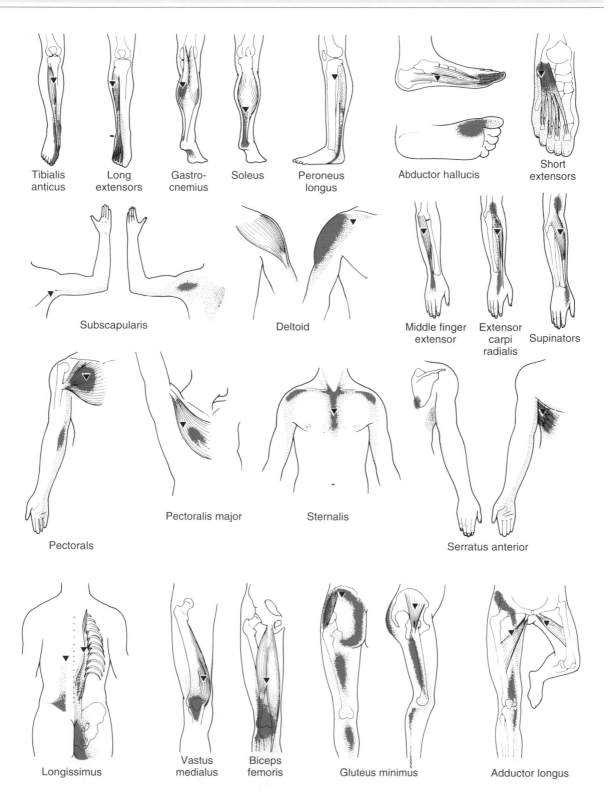

Tibialis anticus

Long extensors

Gastrocnemius

Soleus

Peroneus longus

Abductor hallucis

Short extensors

Subscapularis

Deltoid

Middle finger extensor

Extensor carpi radialis

Supinators

Pectorals

Pectoralis major

Sternalis

Serratus anterior

Longissimus

Vastus medialus

Biceps femoris

Gluteus minimus

Adductor longus

- Trigger points are areas of increased energy consumption and lowered oxygen supply due to inadequate local circulation. They will therefore add to the drain on energy and any fatigue being experienced.
- The muscles in which trigger points lie cannot reach their normal resting length – i.e. they are held in a shortened position.
- Until muscles can reach this normal resting length without pain or effort, any treatment of the trigger points will only achieve temporary relief, since they will reactivate when restressed sufficiently.
- Stretching of the muscles, using either active or passive methods, is useful in treating both the shortness and the trigger point since this can reduce the contraction (taut band) as well as increasing circulation to the area.
- There are many ways of treating trigger points including acupuncture, procaine injections, direct manual ischaemic pressure, stretching, ice therapy, some of which (pressure, acupuncture) cause the release of endorphins, which explains one of the ways in which pain is reduced – another involves the substitution of one sensation (pressure, needle) for another in which pain messages are partially or totally blocked from reaching or being registered by the brain.
- Other treatment methods (stretching for example) alter the dynamics of the circulatory imbalance affecting trigger points and appear to deactivate them.
- The target area to which a trigger refers pain will be the same in everyone if the trigger point is in the same position – but this distribution of pain does not relate directly to neural pathways or to acupuncture meridian pathways.

- The way in which a trigger point relays pain to a distant site is thought to involve one of a variety of neurological mechanisms, and probably involves the brain 'mislocating' pain messages which it receives via several different pathways.
- The sites of trigger points lie in parts of muscles (postural or phasic) most prone to mechanical stress, producing circulatory inadequacy and lack of oxygen – among other changes.
- Trigger points become self-perpetuating (pain–increased tone–pain, cycle) and will never go away unless adequately treated.

INIT – Integrated Neuromuscular Inhibition Technique

The author (Chaitow 1994) has described an integrated sequence in which, after location of an active (referring symptoms) trigger point, this receives ischaemic compression, followed by positional release (osteopathic functional or 'strain/counterstrain' methods), followed (in the same position of ease) by the imposition by the patient of an isometric contraction which is either stretched subsequently (postfacilitation stretch) or simultaneously (isolytic stretch). This combination of methods effectively deactivates trigger points.

Efficient stretching and releasing of whatever soft tissues are short and tight requires an approach which incorporates choices from the variety of different uses of MET as described above. In the following chapters presentation will be made of sequences of assessment and MET treatment methods for the major postural muscles as identified by the work of Lewit, Janda and others, as well as other applications of variations on the theme of MET.

REFERENCES

Bailey M, Dick L 1992 Nociceptive considerations in treating with counterstrain. Journal of the American Osteopathic Association 92: 334–341
Baldry P E 1993 Acupuncture trigger points and musculoskeletal pain. Churchill Livingstone, London
Barlow W 1959 Anxiety and muscle tension pain. British Journal of Clinical Practice 13(5): 339–350

Basmajian J 1978 Muscles alive. Williams and Wilkins, Baltimore
Beal M 1983 Palpatory testing of somatic dysfunction in patients with cardiovascular disease. Journal of the American Osteopathic Association 82: 822–831
Beighton P et al 1983 Hypermobility of joints. Springer Verlag, Berlin

Bennet C 1952 Physics. Barnes and Noble, New York
Block S 1993 Fibromyalgia and the rheumatisms.
Controversies in Clinical Rheumatology 19(1): 61–78
Brookes D 1984 Cranial osteopathy. Thorsons, London
Buller A et al 1960 Interactions between motor neurons and
muscles. J Physiol (London) 150: 417–439
Cantu R, Grodin A 1992 Myofascial manipulation. Aspen
Publications, Maryland
Cathie A 1974 Selected writings. Academy of Applied
Osteopathy Yearbook 1974, pp 15–126
Chaitow L 1991a Palpatory literacy. Thorsons/Harper
Collins, London
Chaitow L 1991b Soft tissue manipulation. Healing Arts
Press, Rochester
Chaitow L 1994 INIT in treatment of pain and trigger
points – an introduction. British Journal of Osteopathy
XIII: 17–21
Cisler T 1994 Whiplash as a total body injury. Journal of the
American Osteopathic Association. 94(2): 145–148
Cyriax J 1962 Textbook of orthopaedic medicine. Cassell
DiGiovanna E 1991 An osteopathic approach to diagnosis
and treatment. Lippincott, Philadelphia
Evjenth O, Hamberg J 1984 Muscle stretching in manual
therapy. Alfta Rehab
Fahrni H 1966 Backache relieved. Thomas Springfield
Garland 1994 Presentation to Respiratory Function
Congress, Paris 1994
Goldenberg D 1993 Fibromyalgia, chronic fatigue syndrome
and myofascial pain syndrome. Current Opinions in
Rheumatology 5: 199–208
Goldthwaite J 1945 Essentials of body mechanics. Lippincott,
Philadelphia
Greenman P 1989 Principles of manual medicine. Williams
and Wilkins, Baltimore
Grieve G 1986 Modern manual therapy. Churchill
Livingstone, London
Gunn C, Milbrandt W 1978 Early and subtle signs in low
back sprain. Spine 3: 267–281
Gutstein R 1955 A review of myodysneuria (fibrositis).
American Practicioner and Digest of Treatments 6(4):
570–577
Guyton A 1987 Basic neuroscience, anatomy and physiology.
W B Saunders, Philadelphia
Isaacson J 1980 Living anatomy – an anatomic basis for
osteopathic theory. Journal of the American Osteopathic
Association 79(12): 752–759
Janda V 1960 Postural and phasic muscles in the
pathogenesis of low back pain. In: Proceedings XI
Congress Rehabilitation International, Dublin
Janda V 1978 Muscles – central nervous motor regulation
and back problems. In: Korr I (ed) Neurobiological
mechanisms in manipulative therapy. Plenum Press,
New York
Janda V 1983 Muscle function testing. Butterworths, London
Janda V 1984 Low back pain – trends, controversies.
Presentation, Turku, Finland, 3–4 September 1984
Janda V 1985 Pain in the locomotor system. In: Glasgow (ed)
Aspects of manipulative therapy. Churchill Livingstone,
London
Janda V 1986a Muscle weakness and inhibition in back pain
syndromes. In: Grieve G (ed) Modern manual therapy of
the vertebral column. Churchill Livingstone, Edinburgh
Janda V 1986b Extracranial causes of facial pain. Journal of
Prosthetic Dentistry 56(4): 484–487

Janda V 1988 In: Grant R (ed) Physical therapy of the cervical
and thoracic spine. Churchill Livingstone, New York
Janda V 1989 Differential diagnosis of muscle tone in respect
of inhibitory techniques. Presentation, PMRF,
21 September 1989
Jones L 1964 Spontaneous release by positioning. The DO
4: 109–116
Jones L 1981 Strain and counterstrain. American Academy of
Osteopathy, Newark
Jones L 1982 Strain and counterstrain. Academy of Applied
Osteopathy, Boulder
Knott M, Voss D 1968 Proprioceptive neuromuscular
facilitation. Hoeber, New York
Komendatov G 1945 Proprioceptivnije reflexi glaza i golovy
u krolikov. Fiziologiceskij Zurnal 31: 62
Korr I 1947 The neural basis of the osteopathic lesion.
Journal of the American Osteopathic Association
48: 191–198
Korr I 1948 The emerging concept of the osteopathic lesion.
Journal of the American Osteopathic Association
48: 127–138
Korr I 1975 Proprioceptors and somatic dysfunction. Journal
of the American Osteopathic Association 74: 638–650
Korr I 1976 Spinal cord as organiser of the disease process.
Academy of Applied Osteopathy Yearbook 1976
Korr I 1978 Sustained sympatheticotonia as a factor in
disease. In: Korr I (ed) The neurobiological mechanisms in
manipulative therapy. Plenum Press, New York
Korr I 1980 Neurobiological mechanisms in manipulation.
Plenum Press, New York
Korr I 1986 Somatic dysfunction, osteopathic manipulative
treatment, and the nervous system. Journal of the
American Osteopathic Association 86(2): 109–114
Korr I et al 1967 Axonal delivery of neuroplasmic
components to muscle cells. Science 155: 342–345
Kraus H 1970 Clinical treatment of back and neck pain.
McGraw Hill, New York
Kuchera M et al 1990 Athletic functional demand and
posture. Journal of the American Osteopathic Association
90(9): 843–844
Lewit K 1974 Functional pathology of the motor system.
Proceedings of the IVth Congress of the International
Federation of Manual Medicine, Prague
Lewit K 1980 Relation of faulty respiration to posture with
clinical implications. Journal of the American Osteopathic
Association 79(8): 525–529
Lewit K 1992a Manipulative therapy in rehabilitation of the
locomotor system. Butterworth/Heinemann
Lewit K 1992b Manipulation in rehabilitation of the motor
system. Butterworth, London
Lewit K, Simons D 1984 Myofascial pain – relief by post-
isometric relaxation. Archives of Physical Medicine and
Rehabilitation 65: 462–466
Liebenson C 1990a Muscular relaxation techniques. Journal
of Manipulative and Physiological Therapeutics 12(6):
446–454
Liebenson C 1990b Active muscular relaxation techniques
(part 2). Journal of Manipulative and Physiological
Therapeutics 13(1): 2–6
Lin J P et al 1994 Physiological maturation of muscles in
childhood. The Lancet June: 1386–1389
McConnell C 1902 Yearbook of Osteopathic Institute of
Applied Technique 1902: 75–78
Mathews P 1981 Muscle spindles. In: Brooks V (ed)

Handbook of physiology. American Physiological Society, Bethseda

Melzack R, Wall P 1988 The challenge of pain. Penguin, New York

Mennell J 1952 The science and art of manipulation. Churchill Livingstone, London

Neuberger A et al 1953 Metabolism of collagen. Biochemistry Journal 53: 47–52

Page L 1952 Academy of Applied Osteopathy Yearbook 1952

Patterson M 1976 Model mechanism for spinal segmental facilitation. Academy of Applied Osteopathy Yearbook 1976

Pauling L 1976 The common cold and 'flu. Freeman, London

Rolf I 1962 Structural dynamics. British Academy of Osteopathy Yearbook 1962

Rolf I 1977 Rolfing – the integration of human structures. Harper and Row, New York

Rothschild B 1991 Fibromyalgia – an explanation. Comprehensive Therapy 17(6): 9–14

Sandman K 1984 Psychophysiological factors in myofascial pain. Journal of Manipulative and Physiological Therapeutics 7(4): 237–242

Scariati P 1991 In: DiGiovanna E An osteopathic approach to diagnosis and treatment. Lippincott, London

Schmid H 1984 Muscular imbalances in skiers. Manual Medicine (2): 23–26

Selye H 1976 The stress of life. McGraw-Hill, New York

Simons D, Travell J 1983 The trigger point manual. Williams and Wilkins, Baltimore

Sommer H 1985 Patellar chodropathy and apicitis – muscle imbalances of the lower extremity. Butterworths, London

Travell J, Simons D 1983 Myofascial pain and dysfunction – the trigger point manual (Vol 1). Williams and Wilkins, Baltimore

Travell J, Simons D 1992 The trigger point manual (Vol 2). Williams and Wilkins, Baltimore

Van Buskirk R 1990 Nociceptive reflexes and the somatic dysfunction. Journal of the American Osteopathic Association 90(9): 792–809

Walther D 1988 Applied kinesiology. SDC Systems, Pueblo

WHO 1981 IIIrd Report on Rehabilitation. WHO, Geneva

3

How to use MET

PALPATION SKILLS

In Chapter 1 a number of variations on the theme of MET were described, as used by clinicians such as Karel Lewit, Vladimir Janda, Craig Liebenson, Aaron Mattes, Edward Stiles and others. In Chapter 4, a sequence is described for the evaluation/assessment of the major postural muscles of the body – for relative shortness – along with details of suggested MET approaches for normalising, stretching and relaxing those muscles.

In this chapter, suggestions are given as to how to begin to learn the application of those MET methods, both for muscles and for joints.

A primary requirement for the operator is the identification, by assessment, of a need for the use of MET; this brings us to a need for sound palpation skills.

EASE AND BIND

The concept and reality of tissues providing palpating hands or fingers with a sense of their relative 'tension' or 'bind', as opposed to their state of 'relaxation' or 'ease', is one which the beginner needs to grasp and the advanced operator probably takes for granted. There can never be enough focus on these two characteristics, which allow the tissues to speak as to their current degree of comfort or distress.

Osteopathic pioneer H V Hoover (1969) describes 'ease' as a state of equilibrium, or 'neutral', which the operator senses by having at least one completely passive 'listening' contact (either the whole hand or a single of several fingers or thumb) in touch with the tissues being assessed. Bind is, of course, the opposite of ease

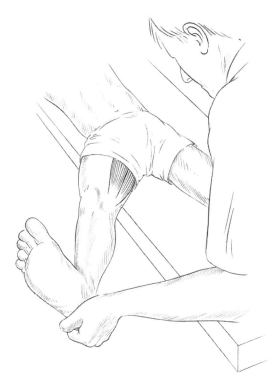

Figure 3.1A Assessment of 'bind'/restriction barrier with the first sign of resistance in the adductors (medial hamstrings) of the right leg. In this example, the operator's perception of the transition point, where easy movement alters to demand some degree of effort, is regarded as the barrier.

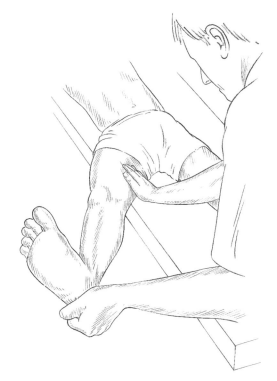

Figure 3.1B Assessment of 'bind'/restriction barrier with the first sign of resistance in the adductors (medial hamstrings) of the right leg. In this example, the barrier is identified when the palpating hand notes a sense of bind in tissues which were relaxed (at ease) up to that point.

and can most easily be noted by lightly palpating the tissues surrounding, or associated with, a joint as this is taken towards the end of its range of movement – its resistance barrier.

In order to 'read' hypertonicity, palpation skills need to be refined, and, as a first step, Goodridge suggests the following test – which examines medial hamstring and short adductor status – as a means of becoming comfortable with the reality of ease and bind in a practical manner (Goodridge 1981).[1]

---

[1]This test, its interpretation and suggested treatment using MET (should shortness be apparent), will be fully explained in Chapter 4, but in this setting we are using it as an exercise for the purposes of becoming familiar with 'ease and bind', and not for actually testing the muscles involved.

Test for palpation of ease and bind during assessment of adductors of the thigh (Fig. 3.1; see also Fig. 1.3 p. 16)

Before testing for ease and bind, a simpler assessment is needed, as described by John Goodridge. He presents a simple method for beginning to become familiar with MET. Before starting ensure that the patient/model lies supine, so that the non-tested leg is abducted slightly, heel over end of table. The leg to be tested is close to the edge of table. (Ensure that the tested leg is in the anatomically correct position, knee in full extension and with no external rotation of the leg, which would negate the test.)

Goodridge's beginners exercise – method

1. 'After grasping the supine patient's foot and ankle, in order to abduct the lower limb, the operator closes his eyes during the abduction, and feels, in his own body, from his hand through his forearm, into his upper arm, the beginning of a sense of resistance.'

2. 'He stops when he feels it, opens his eyes, and notes how many degrees in an arc, the patient's limb has travelled.'

What Goodridge is trying to establish in the individual who is learning MET is that they sense the very beginning of the end of range of free movement, where easy motion ceases and effort on the part of the operator moving the part begins. This 'barrier' is not a pathological one, but represents the first sign of resistance, the place at which tissues require some degree of passive effort to move them. This is also the place at which 'bind' should be palpated.

It is suggested that the process described by Goodridge be attempted several times, so that you get a sense of where resistance begins. Then do the exercise again as described below.

Beginners exercise – variation:

Stand between the patient's partially abducted leg and the table, facing the head of the table, so that all control of the tested leg is achieved by using your lateral arm/hand which holds and supports the leg at the ankle while your table-side hand rests on the inner thigh, palpating the muscles which are being tested. This palpating hand (often called a 'listening' hand in osteopathy) must be in touch with the skin, moulded to the contours of the tissues being assessed, but should exert no pressure, and should be completely relaxed.

Abduction of the tested leg is introduced passively by the outside hand/arm, until the first sign of resistance is noted by the hand which is providing the motive force, i.e. the one holding the leg.

As you approach this point of resistance can you sense a tightening of the tissues in the mid-inner thigh which your table-side hand is touching?

This is bind. If this sensation is not clear then take the leg back towards the table and out again, but this time go past the point where easy movement is lost and effort begins, and towards its end of range. Here you will certainly sense 'bind'. As you once more take the leg back towards the table you will note a softening, a relaxation, an 'ease', in these same tissues. Go through the same sequence with the other leg, becoming increasingly familiar with the sense of these two extremes, and try to note the very moment at which you can palpate the transition from one to the other, not to its extreme but where it begins, whether you are moving from ease to bind or the other way. Normal excursion of the straight leg into abduction is around 45°, and by testing both legs in the manner described you can evaluate whether they are both tight and short, or whether one is and the other is not. Even if both are tight and short one may be more restricted than the other. This is the one to treat first using MET.

It is suggested that before using MET clinically, you practice performing palpation exercises for ease and bind on many other muscles when they are being both actively and passively moved, until you are comfortable with your skill in reading this change in tone. Returning to the exercise suggested by Goodridge, once you feel that you can identify the beginnings of bind in the adductors, and having decided which leg to treat, you can experiment with the simple use of MET.

The point at which you feel bind (or where the hand carrying the leg feels the first sign that effort is required) is the resistance barrier, which will be referred to over and over again in the following chapters; it is the place where an MET isometric contraction can begin in some applications of the methods (notably PIR – see below, p. 50).

It is also the place which is mentally/visually marked if you wish to start from an easier mid-range position, but which you need to note as the place at which resistance was a feature before the isometric contraction.

Identification and appropriate use of the first sign of the barrier of resistance – where bind is first noted – is a fundamental and absolutely critical part of the successful use of MET, along with other key features which include the degree of effort used by the patient and whether you subsequently (after the contraction) take the tissues to a new barrier or through the old one to introduce passive stretching.

Goodridge continues his beginners exercise in MET application as follows (note his parts 1 and 2 above, p. 48):

3. 'The operator compares the arc with the arc produced on the opposite side. In treatment, for example, if the abducted right femur reaches resistance sooner than the left, then restriction of abduction exists. To remove this restriction, the patient's limb is positioned in that arc of movement, where resistance is first perceived, and at this point the operator employs MET to lessen the sense of resistance, and increase the range of movement.'

HOW IS MET USED?

The following exercises in MET variations include the key features emphasised by some of the leading contributors to MET methodology.

Postisometric relaxation (PIR)

Having established the barrier of resistance, either where bind was first noted, or, in Goodridge's words, 'where resistance is first perceived':

- The patient/model is asked to use no more than 20% of their available strength to try to take the leg gently back towards the table – i.e. to adduct the leg – against your firm, unyielding resistance.
- In this example they are trying to take the limb away from the barrier, while you hold it at the barrier.
- The patient/model should be using (contracting) the agonist, the muscle(s) which require release.
- As they induce and hold the contraction they are commonly asked to hold the breath.
- The isometric contraction should be introduced slowly, and resisted without any jerking, wobbling or bouncing.

- Maintaining the resistance to the contraction should produce no strain in the operator.
- The contraction should be held for at least 7, and ideally 10, seconds – the time it is thought necessary for the 'load' on the Golgi tendon organs to become active and to neurologically influence the intrafusal fibres of the muscle spindles, which inhibits muscle tone, so providing the opportunity for the area (muscle, joint) to be taken to a new resting length/resistance barrier without effort, or to stretch it through the barrier of resistance, if this is appropriate (see below) (Scariati 1991).
- The instruction is given, 'Now let your breath go and release your effort, slowly and completely', while you maintain the limb (in this example) at the same barrier.
- The patient/model is asked to breathe in and out once more and to completely relax, and, as he exhales, you gently guide the limb to the point where you now sense a resistance barrier/bind. (You should almost always have increased the range by a significant degree.)
- After use of this method, which induces postisometric relaxation (PIR) in the previously contracted tissues, there exists a latency period of anything from 15 to 30 seconds during which the muscle can be taken to its new resting length, or stretched more easily than would have been the case before the contraction (Guissard et al 1988).

What alternatives have you while in this self-same position?

- You could repeat the exercise precisely as described above, and note whether even more release is possible, working from the resistance barrier to whatever is allowed following the next contraction. This is Lewit's PIR method as described in Chapter 1 (p. 5) and is ideal for releasing tone, for relaxing spasm, for acute conditions.

Janda's approach for chronic/fibrotic tissues

Where fibrosis is a feature, or for chronic conditions, you could use a more vigorous contraction and actually stretch the muscle(s) rather than simply taking them to a new barrier. This would be closer to Janda's approach ('post-facilitation stretch' as described in Ch. 1, p. 7) which calls for the starting of the contraction from a more 'slack', mid-range, position.

- Janda suggests stretching the tissues immediately following cessation of the contraction, and holding the stretch for at least 10 seconds, before allowing a rest period of up to half a minute. Repeat the procedure if necessary.

Modification of Janda's approach

- The recommendation for use of MET for chronic fibrotic tissues, based on osteopathic experience, is that following a contraction of anything up to 20 seconds, which starts in a mid-range position rather than at a barrier, using more than 20% but not more than 50% of the patient's available strength (Janda asks for full strength), a short (2 to 3 seconds) rest period is allowed for complete relaxation before stretch is introduced to a point just beyond the previous barrier of resistance. It is useful to have the patient gently assist in taking the relaxed area towards and through the barrier. The procedure of contraction, relaxation, stretch is repeated (with or without a rest period between contractions) until no more gain is being achieved.

What is different about this use of MET as opposed to Lewit's PIR?

All elements of the procedures as described for PIR are maintained except for:

- starting short of the barrier
- having a longer and stronger contraction, and
- taking the area beyond, rather than just to, the new barrier of resistance (with or without patient assistance).

This procedure is much enhanced by using some patient participation during the stretching procedure – so that they help to take the limb/muscle(s) past the restriction barrier, so minimising the chances of a myotatic stretch reflex being triggered (Mattes 1990).

Janda's approach is undoubtedly successful but carries with it a possibility of very mildly traumatising the tissues (albeit that this is an approach only recommended for chronic and not acute situations). The full-strength contraction which he asks for, and the rapid introduction of stretching following the contraction, are the areas which it is suggested can be most easily modified (as described above) with little loss of successful outcome and with a more secure sense of safety.

Reciprocal Inhibition (RI)

Or, you could try using the alternative physiological mechanism, reciprocal inhibition (RI) which produces a very similar latency ('refractory') period to that produced by PIR.

RI is advocated for acute problems, especially where the muscle(s) requiring release are traumatised, or painful, and cannot easily or safely be used in sustained contractions such as are described in the notes on PIR above.

To use RI according to most guidelines you need to place the area in a 'mid-range' position, short of the resistance barrier (Liebenson 1989). This requirement relates to two factors:

1. The ease of initiating a contraction from a mid-range position as opposed to the relative difficulty of doing so when at an end of range position, and
2. The reduction in risk of inducing cramp which this approach offers, particularly in lower extremity structures such as the hamstrings, especially if longer or stronger contractions than the norm (20% strength, 7 to 10 seconds) are being used.

Staying with the example we have been using you would:

- Test for the sense of bind or increased effort as you abduct the limb, note the degree of excursion of the limb as it reaches this barrier, and then back off a few degrees.
- At this point the patient/model would be asked to try to abduct the leg themselves, taking it towards the barrier, while you

resisted their effort – still asking for no more than 20% of strength.

- Following the end of the contraction (accompanied by breath holding if desired) the same degree of release and relax would be induced, followed by a second breath and more relaxation, at which time you would guide the limb to the new barrier.

Greenman (1989) summarises several of the important component elements of MET as follows:

- There is a patient-active muscle contraction
  — from a controlled position
  — in a specific direction
  — met by operator applied distinct counter-force
  — involving a controlled intensity of contraction.

The common errors which he notes, and which are as important to memorise as are the directions for use of MET, include those listed below.

Patient errors during MET

(Usually based on inadequate instruction from the operator!)

- Contraction is too hard (remedy: give specific guidelines – e.g. use only '20% of strength' or whatever is most appropriate).
- Contraction is in wrong direction (remedy: give simple but accurate instructions).
- Contraction is not sustained for long enough (remedy: instruct the patient to hold the contraction until told to ease off, and give an idea ahead of time as to how long this will be).
- Patient does not relax completely after the contraction (remedy: have them release and relax and then inhale and exhale once or twice with the guide that they should let go completely).

To this list the author would add:

- Starting and/or finishing the contraction too hastily. There should be a slow build-up of force and a slow letting go; this is easily achieved if a rehearsal is carried out first, to educate the patient into the methodology.

Operator errors in application of MET

These include:

- Inaccurate control of position of joint or muscle in relation to the resistance barrier (remedy: have clear image of what is required and apply it).
- Inadequate counterforce to the contraction (remedy: meet and match the force in an isometric contraction, allow movement in an isotonic concentric contraction and overcome the contraction in an isolytic manoeuvre).
- Counterforce is applied in an inappropriate direction (remedy: ensure precise direction needed for best effect).
- Moving to a new position too hastily after the contraction (there is around 25 seconds of refractory muscle tone release during which time a new position can easily be adopted – haste is unnecessary and counter-productive).
- Inadequate patient instruction is given (remedy: get the words right so that the patient can cooperate).

Whenever force is applied, by the patient, in a particular direction, and when it is time to release that effort, the instruction must be to do so gradually. Any quick effort is self-defeating. The coinciding of the forces at the outset (patient and operator) as well as at release is important. The operator must be careful to use enough, but not too much, effort, and to ease off at the same time as the patient.

Contraindications and side-effects of MET

If pathology is suspected no MET should be used until an accurate diagnosis has been established. Pathology (osteoporosis, arthritis etc.) does not rule out the use of MET, but its presence needs to be established so that dosage of application can be modified accordingly (amount of effort used, number of repetitions, stretching introduced or not etc.).

As to side-effects, Greenman explains:

All muscle contractions influence surrounding fascia, connective tissue ground substance and interstitial fluids, and alter muscle physiology by reflex mechanisms. Fascial length and tone is altered by muscle contraction. Alteration in fascia influences not only its biomechanical function, but also its biochemical and immunological functions. The patient's muscle effort requires energy and the metabolic process of muscle contraction results in carbon dioxide, lactic acid and other metabolic waste products which must be transported and metabolised. It is for this reason that the patient will frequently experience some increase in muscle soreness within the first 12 to 36 hours following MET treatment. Muscle energy procedures provide safety for the patient since the activating force is intrinsic and the dosage can be easily controlled by the patient, but it must be remembered that this comes at a price. It is easy for the inexperienced practitioner to overdo these procedures and in essence to overdose the patient.

DiGiovanna (1991) states that side-effects are minimal with MET:

MET is quite safe. Occasionally some muscle stiffness and soreness after treatment. If the area being treated is not localised well or if too much contractive force is used pain may be increased. Sometimes the patient is in too much pain to contract a muscle or may be unable to cooperate with instructions or positioning. In such instances MET may be difficult to apply.

If beginners to MET stay within the very simple guideline which states categorically – cause no pain when using MET – and stick to light (20% of strength) contractions, and do not stretch over-enthusiastically but only take muscles a short way past their restriction barrier when stretching, no side-effects are likely apart from the soreness mentioned above, and this is a normal part of all manual methods of treatment. While the author advocates that this be kept as a guideline for all therapists and practitioners exploring the MET approach, not all texts advocate a completely painless use of stretching and the contrary view needs to be recorded.

Sucher, for example, suggests that discomfort is inevitable with stretching techniques, especially when self-applied at home (Sucher 1990): 'There should be some discomfort, often somewhat intense locally ... however symptoms should subside within seconds or minutes following the stretch.' Kottke (1982) says, 'Stretching should be past the point of pain, but there should be no residual pain when stretching is discontinued.'

Clearly what is noted as pain for one individual will be described as discomfort by another, making this a subjective exercise. Hopefully, sufficient emphasis has been given to the need to keep stretching associated with MET light, just past the restriction barrier, and any discomfort tolerable to the patient.

Breathing and MET

Many of the guidelines for application of isometric contraction call for patient participation over and above their 'muscle energy' activity, most notably involving the holding of a breath during the contraction/effort, and the release of a breath as the new position or stretch is passively or actively adopted. Is there any valid evidence to support this apparently clinically useful element of MET methodology?

There is certainly 'common practice' evidence, for example, in weight training, where the held breath is a feature of the harnessing and focusing of effort, and in yoga practice where the released breath is the time for adoption of new positions. Fascinating as such anecdotal material might be it is necessary to explore the literature for evidence which carries more weight, and fortunately this is available in abundance.

Cummings and Howell (1990) have looked at the influence of respiration on myofascial tension and have clearly demonstrated that there is a mechanical effect of respiration on resting myofascial tissue (using the elbow flexors as the tissue being evaluated).

They also quote the work of Georgiev and Kisselkova who reported that resting EMG activity of the biceps brachii, quadriceps femoris and gastrocnemius muscles, 'cycled with respiration following bicycle ergometer exercise, thus demonstrating that nonrespiratory muscles receive input from the respiratory centres.' (Kisselkova & Georgiev 1976.) The conclusion was that 'these studies document both a mechanically and a neurologically mediated influence on the tension produced by myofascial tissues, which gives objective verification of the clinically observed influence of respiration on the musculoskeletal system and validation of its potential role in manipulative therapy.'

So there is an influence, but what variables does it display? Lewit helps to create subdivisions in the simplistic picture of 'breathing-in enhances effort' and 'breathing-out enhances movement', and a detailed reading of his book *Manipulative Therapy in Rehabilitation of the Locomotor System*, is highly recommended for those who wish to understand the complexities of the mechanisms involved.

Among the simpler connections which he discusses, and for which evidence is provided, are the following facts:

- The abdominal muscles are assisted in their action during exhalation, especially against resistance
- Movement into flexion of the lumbar and cervical spine is assisted by exhalation, and
- Movement into extension (i.e. straightening up, bending backwards) of the lumbar and cervical spine is assisted by inhalation, whereas
- Movement into extension of the thoracic spine is assisted by exhalation (try it and see how much more easily the thoracic spine extends as you exhale than when you inhale) and thoracic flexion is enhanced by inhalation
- Rotation of the trunk in seated position is enhanced by inhalation and inhibited by exhalation
- Neck traction (stretching) is easier during exhalation but lumbar traction (stretching) is eased by inhalation and retarded by exhalation.

Other MET variables

Mitchell's strength testing
Before applying MET to an apparently short muscle, Mitchell suggests (Mitchell et al 1979) that it and its pair should be assessed for relative strength. If the muscle requiring lengthening tests as weaker than its pair, he calls for the reasons for this relative weakness to be evalu-

ated and treated. For example, an antagonist might be inhibiting it and this factor should be dealt with so that the muscle which is due to receive MET attention is strengthened. At this time MET, as described above (pp 50–52), can most suitably be used, according to Mitchell.

Goodridge (1981) concurs and states:

When a left–right asymmetry in range of motion exists, in the extremities that asymmetry may be due to either a hypertonic or hypotonic condition. Differentiation is made by testing for strength, comparing left and right muscle groups. If findings suggest weakness is the cause of asymmetry in range of motion, the appropriate muscle group is treated to bring it to equal strength with its opposite number before range of motion is retested to determine whether shortness in a muscle group may also contribute to the restriction.

The antagonists to any muscle which tests as weak will probably be found to be contracted, and these should first receive attention, using MET. Testing for shortness may give evidence of obvious restriction, and it is questionable whether tests for relative strength are of value in such a scenario. Following MET treatment of whatever is found to be short and/or hypertonic, subsequent assessment may show that hypotonic muscles require toning and this may be achieved using isotonic contractions. Reference to strength testing will be made periodically in descriptions of MET application to particular muscles in Chapter 4, where this factor seems more important clinically, especially in regard to its mention by Mitchell et al.

Is this view valid? Janda in particular gives evidence of the relative lack of accuracy in strength testing, preferring instead an assessment of balanced or unbalanced function and relative shortness in particular structures, considered in the context of overall musculoskeletal function, as a means of assessing what needs attention.

Mitchell and Janda on 'the weakness factor'

Mitchell et al's (1979) recommendation regarding strength testing prior to use of MET complicates the approach advocated by the author, which is to use indications of overactivity or stress, or, even more importantly, signs of mal-coordination and imbalance, as a possible clue to a postural muscle being short (by virtue of it being stressed), along with objective evidence of the same (using one of the many such tests described below, Ch. 4).

Additional evidence of a need to use MET stretching could be derived from basic palpation which indicates the presence of fibrosis and/or myofascial trigger point activity, or of inappropriate electromyographic (EMG) activity, should such technology be available.

Janda effectively dismisses the idea of using strength tests to any degree in evaluating functional imbalances (Janda 1993, Kraus 1970), when he states:

Individual muscle strength testing is unsuitable because it is insufficiently sensitive and does not take into account evaluation of coordinated activity between different muscle groups. In addition in patients with musculoskeletal syndromes, weakness in individual muscles may be indistinct, thus rendering classical muscle testing systems unsatisfactory. This is probably one of the reasons why conflicting results have been reported in studies of patients with back pain.

Also, as noted above (p. 54), Janda tells us that weak muscles, if short, will regain tone if stretched appropriately.

Ideally therefore some observable and/or palpable evidence of functional imbalance will be available which can guide the therapist/practitioner as to the need for MET or other interventions in particular muscles.[2]

For example, in testing for overactivity and, by implication, shortness, in quadratus lumborum there can be an attempt to assess the sequence involved in raising the leg laterally in a sidelying posture. There is a correct and an incorrect (or balanced and unbalanced) sequence, and if the latter is noted stress is proved and, since this is a postural muscle, shortness can be assumed.

The reader must decide whether to introduce Mitchell's element of strength testing into the assessment protocol which they adopt. This recommendation by Mitchell, Moran and Pruzzo will not be detailed in each paired muscle discussed in the text, and is highlighted here (and in a few specific muscles where they lay great emphasis on its importance) in order to remind the reader of the possibility of its incorporation into the methodology of MET use.

The author has not found that application of weakness testing (as part of the work-up before deciding on the suitability or otherwise of MET use for particular muscles) significantly improves results. He does, however, recognise that in individual cases it might be a useful approach and considers that this factor may be left until later in a treatment programme when, after dealing with shortness, attention to weakness may be usefully initiated.

Strength testing methodology

In order to test a muscle for strength a standard procedure is carried out as follows:

- The area should be relaxed and not influenced by gravity
- The area/muscle/joint should be positioned so that whatever movement is to be used can be easily performed
- The patient should be asked to perform a concentric contraction which is evaluated against a scale as given in Box 3.1.

The degree of resistance required to prevent movement is a subjective judgment unless mechanical resistance and/or electronic measurement

---

[2]This topic is discussed further in Chapter 5 which is devoted to Dr Liebenson's views on rehabilitation and which further discusses aspects of Vladimir Janda's functional tests. Some of Janda's, as well as Lewit's, functional assessments are included in the specific muscle evaluations given in Chapter 4.

Box 3.1 Scale for evaluation of concentric contractions

Grade 0 = no contraction/paralysis
Grade 1 = no motion noted but contraction felt by palpating hand
Grade 2 = some movement possible on contraction, if gravity influence eliminated ('poor')
Grade 3 = Motion possible against gravity's influence ('fair')
Grade 4 = Movement possible during contraction against resistance ('good').

is available. For more detailed knowledge of muscle strength testing, texts such as Janda's *Muscle Function Testing* (Butterworths, London 1983) are recommended.

Ruddy's methods – 'pulsed MET'

In the 1940s and 50s, osteopathic physician T. J. Ruddy developed a method of rapid pulsating contractions against resistance which he termed rapid resistive duction. It was in part this work which Fred Mitchell Snr. used as his base for the evolution of MET, along with PNF methodology. Ruddy's method called for a series of muscle contractions against resistance, at a rate a little faster than the pulse rate. This approach can be applied in all areas where isometric contractions are suitable, and is particularly useful for self-treatment following instruction from a skilled practitioner.

Ruddy's work is now known as pulsed MET rather than the tongue-twisting 'Ruddy's rapid resistive duction'. Its simplest use involves the dysfunctional tissue/joint being held at its resistance barrier, at which time the patient, ideally (or the operator if the patient cannot adequately cooperate with the instructions), against the resistance of the operator, introduces a series of rapid (2 per second), minute efforts towards (or sometimes away from) the barrier. The barest initiation of effort is called for with, to use Ruddy's term, 'no wobble and no bounce.'

The use of this 'conditioning' approach involves, in Ruddy's words, contractions which are, 'short, rapid and rhythmic, gradually increasing the amplitude and degree of resistance, thus conditioning the proprioceptive system by rapid movements.'

In describing application of this method to the neck (in a case of vertigo) he gives instruction as to the directions in which the series of resisted efforts should be made. These must include 'movements . . . in a line of each major direction, forwards, backwards, right forward and right backwards or along an antero-posterior line in four directions along the multiplication "X" sign, also a half circle, or rotation right and left.'

His suggested timing is to count each series of contractions as follows: 1–2, 1–2; 2–1, 2–2; 3–1,

3–2; 4–1, 4–2 and so on up to 10–2. These 'mini-contractions' are meant to be timed to coincide with each count so that when complete in all the directions available, an effective series of pulsating contractions against the agonists or antagonists will have been achieved.

If reducing joint restriction or elongation of a soft tissue is the objective then, following each series of 20 contractions of this sort, the slack should be taken out and another series of contractions should be commenced from the new barrier, possibly in a different direction which can and should be varied according to Ruddy's guidelines, to take account of all the different elements in any restriction. Despite Ruddy's suggestion that the amplitude of the contractions be increased over time, the effort itself must never exceed the barest beginning of an isometric contraction.

The effects are likely, Ruddy suggests, to include improved oxygenation, and improved venous and lymphatic circulation through the area being treated. Furthermore, he believes that the method influences both static and kinetic posture because of the effects on proprioceptive and interoceptive afferent pathways, and that this helps maintain 'dynamic equilibrium', which involves 'a balance in chemical, physical, thermal, electrical and tissue fluid homeostasis.'

Ruddy's work was a part of the base on which Mitchell and others constructed MET and his work is worthy of study and application since it offers, at the very least, a useful means of modifying the employment of sustained isometric contraction usage, which has particular relevance to acute problems and safe self-treatment settings.

Isotonic concentric strengthening methods

1. The operator positions the limb, or area, so that the muscle group will be at resting length, and thus will develop the strongest contraction.
2. The operator explains the direction of movement required, as well as the intensity and duration of that effort. The patient contracts the muscle with the objective of moving the muscle through a complete range, quickly (about 2 seconds).

3. The operator offers counterforce, which is less than that of the patient's contraction, and maintains this throughout the contraction. This is repeated several times, with a progressive increase in operator's counterforce (the patient's effort in the strengthening mode is always maximal).

Where weak muscles are being toned, via isotonic methods, the operator allows the concentric contraction of the muscles, as the patient attempts to move in a manner which employs the hypotonic structures. There are several methods possible. Such exercises always involve an operator's force which is less than that applied by the patient. The subsequent isotonic concentric contraction of the weakened muscles allows approximation of the origins and insertions to be achieved under some degree of control by the operator. In such cases the efforts are usually suggested as being of short duration, ultimately employing maximal effort on the part of the patient.

This use of concentric isotonic contractions to tone a muscle or muscle group can be expanded to become an isokinetic, whole joint movement.

Strengthening a joint complex with isokinetic MET

The major variation on this method of simple isotonic contraction is to use what has been called isokinetic contraction (also known as progressive resisted exercise). In this the patient, starting with a weak effort, but rapidly progressing to a maximal contraction of the affected muscle(s), introduces a degree of resistance to the operator's effort to put a joint, or area, through a full range of motion. An alternative is for the operator to partially resist the patient's active movement of a joint through a rapid series of as full a range of movements as possible.

Mitchell describes an isokinetic exercise as follows: 'the counterforce is increased during the contraction to meet changing contractile force as the muscle shortens and its force increases.' These are, he says, especially valuable in improving efficient and coordinated use of muscles and of enhancing the tonus of the resting muscle. 'In dealing with paretic muscles,

isotonics (in the form of progressive resistance exercise) and isokinetics are the quickest and most efficient road to rehabilitation.'

The use of isokinetic contraction is reported to be a most effective method of building strength, and to be superior to high repetition, lower resistance exercises (Blood 1980). It is also felt that a limited range of motion, with good muscle tone, is preferable (to the patient) to normal range with limited power. Thus the strengthening of weak musculature in areas of permanent limitation of mobility is seen as an important contribution in which isokinetic contractions may assist.

Isokinetic contractions not only strengthen the fibres which are involved, but have a training effect which enables them to operate in a more coordinated manner. There is often a very rapid increase in strength. Because of neuromuscular recruitment, there is a progressively stronger muscular effort as this method is repeated. Contractions, and accompanying mobilisation of the region, should take no more than 4 seconds at each contraction in order to achieve maximum benefit with as little fatiguing as possible of either the patient or the operator. Prolonged contractions should be avoided. The simple and safest applications of isokinetic methods involve small joints, such as those in the extremities. Spinal joints may be more difficult to mobilise whilst muscular resistance is being fully applied. The options in achieving increased strength, via these methods, therefore involves a choice between a partially resisted isotonic contraction or the overcoming of such a contraction, at the same time as the full range of movement is being introduced. Both of these options can involve maximum contraction of the muscles by the patient. Home treatment of such conditions is possible, via self-treatment, as in other MET methods.[3]

DiGiovanna (1991) suggests that isokinetic exercise increases the work a muscle can perform more efficiently and rapidly than either isometric or isotonic exercises.

---

[3]Both isotonic concentric and eccentric contractions, will take place during the isokinetic movement of a joint.

Reduction of fibrotic changes with isolytic MET

Another application of isotonic contraction is that in which the patient initiated contraction is resisted and overcome by the operator. This is an eccentric contraction, in that the origins and insertions of the muscles involved will become further separated, despite the patient's effort to approximate them, and has been termed isolytic contraction, in that it involves the stretching, and sometimes the breaking down, of fibrotic tissue present in the affected muscles. Micro-trauma is inevitable and this form of 'controlled' injury is seen to be useful especially in relation to altering the interface between elastic and non-elastic tissues – between fibrous and non-fibrous tissues.

Mitchell states, 'advanced myofascial fibrosis sometimes requires this "drastic" measure, for it is a powerful stretching technique.'

Adhesions of this type are broken down by the application of force by the operator which is just a little greater than that of the patient. This procedure can be uncomfortable, and the patient should be advised of this, as well as of the fact that they need only apply sufficient effort to ensure that they remain comfortable. Limited degrees of effort are therefore called for at the outset of isolytic contractions.

However, in order to achieve the greatest degree of stretch, in the condition of myofascial fibrosis for example, it is necessary for the largest number of fibres possible to be involved in the isotonic contraction. Thus there is a con-tradiction in that, in order to achieve this large involvement, the degree of contraction should be a maximal one, which could produce pain, which is contraindicated.

Additionally, in many instances, the procedure might be impossible to achieve if a large muscle (hamstrings) group is involved and were a maximal contraction to be used, especially if the patient is strong and the operator slight or at least inadequate to the task of overcoming the contracting muscle(s). Less than optimal con-traction is therefore called for, repeated several times perhaps, but confined to specific muscles where fibrotic change is greatest (tensor fascia lata [TFL], for example) and to patients who are not frail, pain-sensitive or in other ways unsuitable for what is the most vigorous MET method.

Summary of choices for MET in treating muscle problems

To return to Goodridge's introduction to MET (see pp 48–49) – using the adductors as our target tissues, we can now see that a number of choices are open to you once you have established that the objective is to lengthen shortened adductor muscles; for example, were the objective to lengthen shortened adductors, on the right, several methods could be used:

- The patient could contract the right abductors, against equal operator counterforce, in order to relax the adductors by reciprocal inhibition; or
- The patient could contract the right adductors against equal operator counterforce, in order to achieve postisometric relaxation; or
- The patient could contract the right adductors whilst the operator offered greater counterforce, thus overcoming the isotonic contraction (eccentric isotonic, or isolytic, contraction).

In all these examples the shortened muscles would have been taken to their appropriate barrier before commencing the contraction – either at the first sign of resistance if PIR and movement to a new barrier was the objective, or in a mid-range position if RI or a degree of postfacilitation stretching was considered more appropriate. For an isolytic stretch the contrac-tion commences from the resistance barrier, as do all isokinetic activities.

The essence of muscle energy methods then is the harnessing of the patient's own muscle power. The next prerequisite is the application of counterforce in a predetermined manner. In isometric methods this counterforce must be unyielding. No test of strength must ever be attempted. Thus the patient should never be asked to 'try as hard as he can' to move in this or that direction. It is important before commen-cing that this, and the rest of the procedure, be carefully explained, so that the patient has a clear idea of his role. The direction, limited degree of effort, and duration, must be clear, as must the associated instructions regarding breathing patterns, and eye movements (if any).

Joints and MET

MET uses muscles and soft tissues for its effects; nevertheless, the impact of these methods on joints is clearly profound since it is impossible to consider joints independent of muscles. For practical purposes, however, an artificial division is made in the text of this book and in Chapter 6 there will be specific focus given to topics such as MET in treatment of joint restriction and dysfunction; preparing joints for manipulation with MET; as well as the vexed question of the primacy of muscles or joints in dysfunctional settings. The opinions of experts such as Hartman, Stiles, Evjenth, Lewit, Janda, Goodridge and Harakal will be outlined in relation to these and other joint related topics.

A chiropractic view is provided in Chapter 5, which looks at rehabilitation implications of MET as its prime interest but which touches on the treatment protocol which a chiropractic expert, Craig Liebenson, suggests in relation to dysfunctional imbalances which involve joint restriction/blockage.

Self-treatment

Lewit (1991) is keen to involve patients in home treatment, using MET. He describes this aspect thus:

Receptive patients are taught how to apply this treatment to themselves, as autotherapy, in a home programme. They passively stretched the tight muscle with their own hand. This hand next provided counter pressure to voluntary contraction of the tight muscle (during inhalation) and then held the muscle from shortening, during the relaxation phase. Finally, it supplied the increment in range of motion (during exhalation) by taking up any slack that had developed.

How often should self-treatment be prescribed?

Gunnari (Gunnari & Evjenth 1983) recommends frequent application of mild stretching or, if this is not possible, then more intense but less frequent self-stretching at home. He states,

'Therapy is more effective if it is supplemented by more frequent self-stretching. In general, the more frequent the stretching, the more moderate the intensity; less frequent stretching, such as that done every other day, may be of greater intensity.'

Self-treatment methods are not suitable to all regions (or to all patients) but there are a large number of areas which lend themselves to such methods. Use of gravity, as a counter pressure source, is often possible in self-treatment. For example, in order to stretch quadratus lumborum (see Fig. 3.2), the patient stands, legs apart and sidebending, in order to impose a degree of stretch to the shortened muscle. By inhaling and slightly pushing the trunk towards an upright position, against the weight of the trunk, which gravity is pulling towards the floor, and then releasing the breath at the same time as trying to bend further towards the side, a greater degree of movement will have been achieved.

Lewit suggests, in such a movement, that the counter-movement against gravity be accompanied by movement of the eyes upwards, and the attempt to bend further to the side, by looking downward. These eye movements facilitate the effects. Several attempts by the patient to induce greater freedom of movement, in any restricted direction, should achieve good results by means of such simple measures.

The principles of MET are now hopefully clearer and the methods seen to be applicable to a large range of problems.

Rehabilitation, as well as first-aid, and some degree of normalisation of both acute and chronic soft-tissue and joint problems, are all possible, given correct application. Combined with NMT, this offers the practitioner the chance of achieving safe and effective therapeutic intervention.

WHEN SHOULD MET (PIR, RI OR POSTFACILITATION STRETCH) BE APPLIED TO A MUSCLE TO RELAX AND/OR STRETCH IT?

1. When it is demonstrably shortened – unless the shortening is attributable to associated joint restriction, in which case this should

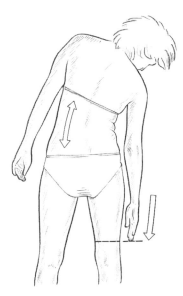 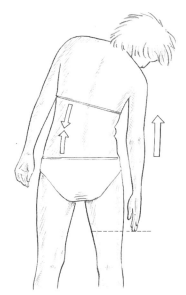 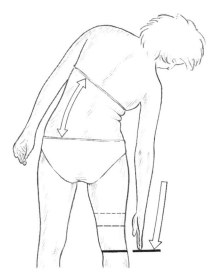

Figure 3.2 MET self-treatment for quadratus lumborum.
Figure 3.2A Patient assesses range of sidebending to the right.

Figure 3.2B Patient contracts quadratus lumborum by straightening slightly, thereby introducing an isometric contraction against gravity.

Figure 3.2C After 7–10 seconds, the contraction is released and the patient will be able to sidebend further, stretching quadratus lumborum towards its normal resting length.

receive primary attention, possibly also involving MET (see Ch. 6, p. 114).

2. When it contains areas of shortening, such as myofascial trigger points or palpable fibrosis. It is important to note that trigger points evolve within stressed (hypertonic) areas of phasic as well as postural muscles, and that these will require stretching, based on evidence which shows that trigger points reactivate unless shortened fibres in which they are housed are stretched to a normal resting length as part of a treatment plan (Travell & Simons 1983).

3. When periosteal pain points are palpable, indicating stress at the associated muscle's origin and/or insertion.

4. In cases of muscular imbalance in order to reduce hypertonicity when weakness in a muscle is attributable, in part or totally, to inhibition deriving from a hypertonic antagonist muscle (group).

Evaluation

It is seldom possible to totally isolate one muscle in an assessment, and reasons other than muscle shortness can account for apparent restriction (intrinsic joint dysfunction for example).

Other methods of evaluation as to relative muscle shortness are also called for, including direct palpation.

The 'normal' range of movements of particular muscles should be taken as guidelines only, since individual factors will often determine that what is 'normal' for one is not so for another.

Wherever possible, an understanding is called for of functional patterns which are observable, for example in the case of the upper fixators of the shoulder/accessory breathing muscles. If a pattern of breathing is observed which indicates a predominance of upper chest involvement, as opposed to diaphragmatic, this in itself would indicate that this muscle group was being 'stressed' by overuse. Since stressed postural muscles will shorten, an automatic assumption of shortness can be made in such a case regarding the scalenes, levator scapulae etc. (see Ch. 2, p. 32 for a fuller discussion of Janda's evidence for this and p. 34 for Garland's description of structural changes relating to this pattern of breathing).

Once again let it be clear that the various tests and assessment methods suggested in Chapter 4, even when utilising evidence of an abnormally short range of motion, are meant as indicators of, and not certainties, as to shortness (Gunnari & Evjenth 1983).

As Evjenth observes, 'If the preliminary analysis identifies shortened muscles, then a provisional trial treatment is performed. If the provisional treatment reduces pain and improves the affected movement pattern, the preliminary analysis is confirmed, and treatment may proceed.'

MUSCLE ENERGY TECHNIQUE: SUMMARY OF VARIATIONS

1. Isometric contraction – using reciprocal inhibition

Indications

*   Relaxing muscular spasm or contraction
*   Mobilising restricted joints
*   Preparing joint for manipulation.

Contraction starting point At the restriction barrier.

Modus operandi Affected muscle(s) not employed, antagonists used in isometric contraction obliging shortened muscles to relax via reciprocal inhibition. Patient is attempting to push through the barrier of restriction against operator's precisely matched counterforce.

Forces Operator's and patient's forces are matched. Initial effort involves approximately 20% of patient's strength; an increase to no more than 50% on subsequent contractions is appropriate. Increase of the duration of the contraction – up to 20 seconds – may be more effective than any increase in force.

Duration of contraction 7 to 10 seconds initially, increasing to up to 20 seconds in subsequent contractions, if greater effect required.

Action following contraction Area (muscle/joint) is taken to its new restriction barrier without stretch after ensuring complete relaxation. Perform movement to new barrier on an exhalation.

Repetitions 3 to 5 times or until no further gain in range of motion is possible.

2. Isometric contraction – using postisometric relaxation (without stretching)

Indications

*   Relaxing muscular spasm or contraction
*   Mobilising restricted joints
*   Preparing joint for manipulation.

Contraction starting point At resistance barrier.

Modus operandi Affected muscles (agonists) are used in the isometric contraction, therefore the shortened muscles subsequently relax via postisometric relaxation. Operator is attempting to push through barrier of restriction against the patient's precisely matched counter-effort.

Forces Operator's and patient's forces are matched. Initial effort involves approximately 20% of patient's strength; an increase to no more than 50% on subsequent contractions is appropriate. Increase of the duration of the contraction – up to 20 seconds – may be more effective than any increase in force.

Duration of contraction 7 to 10 seconds initially, increasing to up to 20 seconds in subsequent contractions, if greater effect required.

Action following contraction Area (muscle/joint) is taken to its new restriction barrier without stretch after ensuring patient has completely relaxed. Perform movement to new barrier on an exhalation.

Repetitions 3 to 5 times or until no further gain in range of motion is possible.

3. Isometric contraction – using postisometric relaxation (with stretching, also known as postfacilitation stretching)

Indications

*   Stretching restricted, fibrotic, contracted soft tissues (fascia, muscle).

Contraction starting point Short of resistance barrier, in mid-range.

Modus operandi Affected muscles (agonists) are used in the isometric contraction, therefore the shortened muscles subsequently relax via postisometric relaxation allowing an easier stretch to be performed. Operator is attempting to push through barrier of restriction against the patient's precisely matched counter-effort.

Forces Operator's and patient's forces are matched. Initial effort involves approximately 30% of patient's strength; an increase to no more than 50% on subsequent contractions is appropriate. Increase of the duration of the contraction – up to 20 seconds – may be more effective than any increase in force.

Duration of contraction 7 to 10 seconds initially, increasing to up to 20 seconds in subsequent contractions, if greater effect required.

Action following contraction Rest period of 5 to 10 seconds to ensure complete relaxation before stretch is useful. On an exhalation the area (muscle/joint) is taken to its new restriction barrier and a small degree beyond, painlessly, and held in this position for at least 10 seconds. The patient can participate in helping move the area to and through the barrier, effectively further inhibiting the structure being stretched and retarding the likelihood of a myotatic stretch reflex occurrence.

Repetitions 3 to 5 times or until no further gain in range of motion is possible.

4. Isotonic concentric contraction

Indications

- Toning weakened musculature.

Contraction starting point In a mid-range easy position.

Modus operandi The contracting muscle is allowed to do so, with some (constant) resistance from the operator.

Forces The patient's effort overcomes that of the operator since patient's force is greater than operator resistance. Patient uses maximal effort available, but force is built slowly not via sudden effort. Operator maintains constant degree of resistance.

Duration 3 to 4 seconds.

Repetitions 5 to 7 times or more if appropriate.

5. Isotonic eccentric contraction (isolytic)

Indications

- Stretching tight, fibrotic musculature.

Contraction starting point A little short of restriction barrier.

Modus operandi The muscle to be stretched is contracted and is prevented from doing so by the operator, via superior operator effort, and the contraction is overcome and reversed, so that a contracting muscle is stretched. Origin and insertion do not approximate. Muscle is stretched to, or as close as possible to, full physiological resting length.

Forces Operator's force is greater than patient's. Less than maximal patient's force is employed at first. Subsequent contractions build towards this, if discomfort is not excessive.

Duration of contraction 2 to 4 seconds.

Repetitions 3 to 5 times if discomfort is not excessive.

CAUTION

Avoid using isolytic contractions on head/neck muscles or at all if patient is frail, very pain-sensitive, osteoporotic.

6. Isokinetic (combined isotonic and isometric contractions)

Indications

- Toning weakened musculature
- Building strength in all muscles involved in particular joint function
- Training and balancing effect on muscle fibres.

Starting point of contraction Easy mid-range position.

Modus operandi Patient resists with moderate and variable effort at first, progressing to maximal effort subsequently, as operator puts joint rapidly through as full a range of movements as possible. This approach differs from a simple isotonic exercise by virtue of whole ranges of motion, rather than single motions being involved, and because resistance varies, progressively increasing, as the procedure progresses.

Forces Operator's force overcomes patient's effort to prevent movement. First movements (taking an ankle, say, into all its directions of motion) involves moderate force, progressing to full force subsequently. An alternative is to have the operator (or machine) resist the patient's effort to make all the movements.

Duration of contraction Up to 4 seconds.

Repetitions 2 to 4 times.

REFERENCES

Blood S 1980 Treatment of the sprained ankle. Journal of the American Osteopathic Association

Cummings J, Howell J 1990 The role of respiration in the tension production of myofascial tissues. Journal of the American Osteopathic Association 90(9): 842

DiGiovanna E 1991 Treatment of the spine. In: Osteopathic approach to diagnosis and treatment. Lippincott, Philadelphia

Goodridge J 1981 MET – definition, explanation, methods of procedure. Journal of the American Osteopathic Association 81(4): 249

Greenman P 1989 Principles of manual medicine. Williams and Wilkins, Baltimore

Guissard N et al 1988 Muscle stretching and motorneurone excitability. European Journal of Applied Physiology 58: 47–52

Gunnari H, Evjenth O 1983 Sequence exercise. Norwegian edn. Dreyers Verlag, Oslo

Hoover H 1969 A method for teaching functional technique. Yearbook of Academy of Applied Osteopathy 1969

Janda V 1993 Assessment and treatment of impaired movement patterns and motor recruitment. Presentation to Physical Medicine Research Foundation, Montreal, October 9–11, 1993

Journal of the American Osteopathic Association 1980 79(11): 689

Kisselkova, Georgiev J 1976 Journal of Applied Physiology 46: 1093–1095

Kottke F 1982 Therapeutic exercise to maintain mobility. In: Krusen's handbook of physical medicine and rehabilitation, 3rd edn. W B Saunders, Philadelphia

Kraus H 1970 Clinical treatment of back and neck pain. McGraw Hill, New York

Lewit K 1991 Manipulative therapy in rehabilitation of the locomotor system. Butterworths, London

Liebenson C 1989 Active muscular relaxation methods. Journal of Manipulative and Physiological Therapeutics 12(6): 446–451

Mattes A 1990 Active and assisted stretching. Mattes, Sarasota

Mitchell F, Moran P, Pruzzo N 1979 An evaluation and treatment manual of osteopathic muscle energy technique. Valley Park, Missouri

Ruddy T J 1962 Osteopathic rhythmic resistive technic. Academy of Applied Osteopathy Yearbook 1962: 23–31

Travell J, Simons D 1983 Trigger point manual. Williams and Wilkins, Baltimore

Scariati P 1991 Neurophysiology relevant to osteopathic manipulation. In: DiGiovanna E (ed) Osteopathic approach to diagnosis and treatment. Lippincott, Philadelphia

Sucher B 1990 Thoracic outlet syndrome – a myofascial variant (part 2). Journal of the American Osteopathic Association 90(9): 810–823

4

Sequential assessment and treatment of main postural muscles

Lewit (1985a) summarises what manipulation is concerned with as 'restricted mobility', with or without pain. Evjenth (1984) is equally succinct, and states that what is needed to become proficient in treating patients with symptoms of pain or 'constrained movement' is 'experience gained by thoroughly examining every patient'. The only real measure of successful treatment is, he states, 'restoration of muscle's normal pattern of movement with freedom from pain.'

Janda (1988), however, seems more concerned with 'imbalances' and the implications of dysfunctional patterns in which some muscles become weaker and others progressively tighter.

EVALUATING MUSCLE SHORTNESS

Most of the problems of the musculoskeletal system seem to involve pain related to aspects of muscle shortening. Where weakness (lack of tone) is found to be a major element, it will often be found that antagonists to these muscles are shortened, reciprocally inhibiting their tone. Prior to any effort to strengthen weak muscles, shortened ones should be dealt with by appropriate means, after which spontaneous toning occurs in the previously flaccid muscles. If tone is still inadequate, then, and only then, should exercise and/or isotonic procedures be brought in. Janda tells us that tight muscles usually maintain their strength; however, in extreme cases of tightness some decrease in strength occurs. In such cases stretching (MET) of the tight muscle usually leads to a rapid recovery of strength (as well as toning of their antagonists via removal of reciprocal inhibition).

It is therefore important that we learn to assess short, tight muscles in a standardised manner.

Janda suggests that, in order to obtain a reliable evaluation of muscle shortness:

- The starting position, method of fixation and direction of movement must be observed carefully.
- The prime mover must not be exposed to external pressure.
- If possible, the force exerted on the tested muscle must not work over two joints.
- The examiner should perform, at an even speed, a slow movement that brakes slowly at the end of the range.
- To keep the stretch and the muscle irritability about equal, the movement must not be jerky.
- Pressure or pull must always act in the required direction of movement.
- Muscle shortening can only be correctly evaluated if the joint range is not decreased, as might be the case should an osseous limitation or joint blockage exist (Janda 1983).

It is in shortened muscles, as a rule, that reflex activity is noted. This takes the form of local dysfunction – variously called trigger points, tender points, zones of irritability, neurovascular and neurolymphatic reflexes etc. (Chaitow 1991). Localising these is possible via normal palpatory methods. Identification and treatment of tight muscles may also be systematically carried out using the methods described later in this chapter.[1]

IMPORTANT NOTES ON ASSESSMENTS AND USE OF MET

1. When the term 'restriction barrier' is used in relation to soft tissue structures, it is meant to indicate the place where you note the first signs of resistance (as palpated by sense of 'bind', or sense of effort required to move the area, or by visual or other palpable evidence) and not the greatest possible range of movement obtainable.
2. In all treatment descriptions involving MET

(apart from the first set of assessment tests involving gastrocnemius and soleus) it will be assumed that the 'shorthand' reference to 'acute' and 'chronic' will be adequate to alert the reader to the variations in methodology which these variants call for (see Box 4.1).

3. Assistance from the patient is valuable as movement is made to or through a barrier, provided that the patient can be educated to gentle cooperation and learn not to use excessive effort.
4. In most MET treatment guidelines in this chapter the method described will involve isometric contraction of the agonist(s), that is, the muscle(s) which require stretching. It is assumed that the reader is now familiar with the possibility of using the antagonists to achieve reciprocal inhibition (RI) before initiating stretch or movement to a new barrier and will use this alternative when appropriate (if there is pain on use of agonist; if there has been prior trauma to agonist; in an attempt to see if more release can be made available after the initial use of the agonist isometrically).
5. Isolytic (eccentric isotonic) contractions will be suggested in a few instances, most notably in treating tensor fascia lata (TFL), but these are not generally recommended for application in sensitive patients, or in potentially 'fragile' areas, such as in muscles associated with the cervical spine.
6. Careful reading of Chapters 1 and 3 in particular is urged before commencing practice of the methods listed below.

Box 4.1 'Acute' and 'chronic'

'Acute' and 'chronic' alert the reader to the differences in methodology which these variants call for, especially in terms of the starting position for contractions – acute starts at the barrier, chronic starts short of the barrier – and the operator takes the area to (acute), or through (chronic), the resistance barrier, subsequent to the contraction.

The term 'acute' may be applied to straining which occurred within the past 3 weeks, or where the symptoms are acute.

---

[1]Note that the assessment methods presented are not themselves diagnostic but provide strong indications of probable shortness of the muscles being tested.

7. There should be no pain experienced during application of MET, although mild discomfort (stretching) is acceptable.

8. The methods of assessment and treatment of these postural muscles given here are far from comprehensive or definitive. There are many other assessment approaches, and numerous treatment/stretch approaches, using variations on the theme of MET, as evidenced by the excellent texts by Janda, Basmajian, Lewit, Liebenson, Greenman, Grieve, Mattes, Hartman, Evjenth and Dvorak, among others. The methods recommended below provide a sound basis for the application of MET to specific muscles and areas, as do the methods suggested for spinal, pelvic, neck and shoulder regions in following chapters (Chs 5&6). By developing the skills with which to apply the methods, as described, a repertoire of techniques can be acquired, offering a wide base of choices that will be appropriate in numerous clinical settings.

9. Some of the discussion of particular muscles will include 'notes' containing information unrelated to the main objective, which is to outline assessment and MET treatment possibilities. These notes are included where the particular information they carry is likely to be useful clinically.

10. Breathing cooperation can, and should, be used as part of the methodology of MET. This, however, will not be repeated as an instruction in each example of MET use below. Basically, if appropriate (that is, if the patient is cooperative and capable of following instructions), the patient should be given the instructions outlined in Box 4.2. A note which gives the instruction to 'use appropriate breathing', or some variation on it, will be found in the following text describing various MET applications, and this refers to the guidelines outlined in Box 4.2.

11. Various eye movements are sometimes advocated during contractions and stretches, particularly by Lewit who uses these methods to great effect. The only specific

---

> **Box 4.2** Notes on breathing during MET
>
> If the patient is cooperative and capable of following instructions, they should be asked to:
> - Inhale as they slowly build up an isometric contraction
> - Hold the breath for the 7 to 10 second contraction, and
> - Release the breath on slowly ceasing the contraction
> - They should be asked to inhale and exhale fully once more following cessation of all effort, as they are instructed to 'let go completely'
> - During this last exhalation the new barrier is engaged (acute) or the barrier is passed as the muscle is stretched (chronic).

---

recommendations will be found in regard to muscles such as the scalenes and sternomastoid, where the use of eye movements is particularly valuable in terms of the gentleness of the contractions they induce.

12. 'Pulsed muscle energy technique' is based on Ruddy's work (see Chapter 3, p. 56) and can be substituted for any of the methods described in the text below for treating shortened soft tissue structures, or for increasing the range of motion in joints (Ruddy 1962).

13. There are times when 'co-contraction' is useful, involving contraction of both the agonist and the antagonist. Studies have shown that this approach is particularly useful in treatment of the hamstrings, when both these and the quadriceps are isometrically contracted prior to stretch (Moore at al 1980).

14. It is seldom necessary to treat all shortened muscles which are identified via the methods described below. For example, Lewit and Simons mention that isometric relaxation of the suboccipital muscles will also relax the sternocleidomastoid muscles; treatment of the thoracolumbar muscles induces relaxation of iliopsoas, and vice versa; treatment (MET) of the sternocleidomastoid and scalene muscles relaxes the pectorals. These interactions are worthy of greater study.

Postural muscle assessment sequence checklist

The checklist in Box 4.3 can be used to follow (and record results of) the simple sequence of postural muscle assessment as described in detail later in this chapter.

SEQUENTIAL ASSESSMENT AND MET TREATMENT OF POSTURAL MUSCLES

These assessment and treatment recommendations represent a synthesis of information derived from personal clinical experience and from the numerous sources which are cited, or are based on the work of researchers, clinicians and therapists who are named (Basmajian 1974, Cailliet 1962, Dvorak & Dvorak 1984, Fryette 1954, Greenman 1989, Janda 1983, Lewit 1992, Mennell 1964, Rolf 1977, Williams 1965).[2]

1. Assessment and treatment of gastrocnemius and soleus

Assessment of tight gastrocnemius (01) and soleus (02) (Fig. 4.1)

Patient is supine with feet extending over edge of couch. For right leg examination operator's left hand grasps Achilles tendon just above heel, with no pressure on tendons. The heel lies in the palm of the hand, fingers curving round it. The right hand is placed so that the fingers rest on the dorsum of the foot (these must rest all the time, not apply a pulling stretch) with the thumb on the sole, lying along the medial margin. This position is important as mistakes may involve placing the thumb too near the centre of the sole of the foot. Stretch is introduced by a pull on the heel with the left hand, whilst the right hand maintains the upward

---

Box 4.3 Postural Muscle Assessment Sequence

NAME _____

E = Equal (circle both if both are short)

L & R are circled if left or right are short

Spinal abbreviations indicate low lumbar, lumbodorsal junction, low-thoracic, mid-thoracic and upper thoracic areas (of flatness and therefore reduced ability to flex – short erector spinae)

01. Gastrocnemius E L R
02. Soleus E L R
03. Medial hamstrings E L R
04. Short adductors E L R
05. Rectus femoris E L R
06. Psoas E L R
07. Hamstrings
    a) upper fibres E L R
    b) lower fibres E L R
08. Tensor fascia lata E L R
09. Piriformis E L R
10. Quadratus lumborum E L R
11. Pectoralis major E L R
12. Latissimus dorsi E L R
13. Upper trapezius E L R
14. Scalenes E L R
15. Sternocleidomastoid E L R
16. Levator scapulae E L R
17. Infraspinatus E L R
18. Subscapularis E L R
19. Supraspinatus E L R
20. Flexors of the arm E L R
21. Spinal flattening:
    a) seated legs straight LL LDJ LT MT UT
    b) seated legs flexed LL LDJ LT MT UT
22. Cervical spine extensors short? Yes No

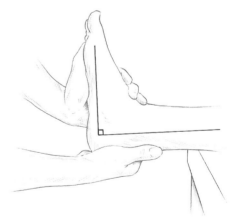

Figure 4.1A Assessment of gastrocnemius and soleus. The sole of the foot should achieve a vertical position without effort once slack is taken out via traction on the heel.

---

[2] The 'code' number assigned to each muscle links it to the Postural Muscle Assessment sequence checklist in Box 4.3.

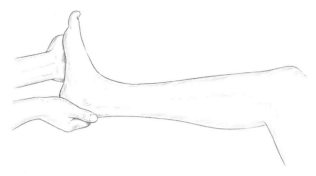

Figure 4.1B With the knee flexed, the same assessment is evaluating the status of soleus alone.

pressure via the thumb (along its entire length). The heel of the right hand prevents sideways movement of the foot.

A range should be achieved which takes the foot to a 90° angle to the leg without any force being applied; it is possible to use the hand which has removed slack from the muscles via traction to note a sense of bind as the foot is dorsiflexed. The leg must remain resting on the couch all the while and the right hand holding/ palpating the muscle insertion and the heel must be placed so that it is an extension of the leg, not allowing an upward (towards the ceiling) pull when stretch is introduced.

An alternative method is to have the patient seated on the couch, legs outstretched, and to have her bend towards the toes with arms extended. If toe touching is possible, but toes are plantarflexed, then there is probably shortness of the gastrocnemius–soleus muscles.

Assessment of tight soleus (02)

The method described above assesses both gastrocnemius and soleus. To assess only the soleus, precisely the same procedure is adopted, with the knee passively flexed (over a cushion, for example). Again, if the foot fails to easily come to a 90° angle without force, once slack has been taken out of the tissues via traction through the long axis of the calf from the heel, soleus is considered short.

A screening test for soleus involves the patient being asked to squat, trunk slightly flexed, feet shoulder width apart, so that the buttocks rest between the legs which face forwards rather than outwards. If soleus are normal then it should be possible to go fully into this position with the heels remaining flat on the floor. If not, and the heels rise from the floor as the squat is performed, soleus muscles are probably shortened.

MET treatment of shortened gastrocnemius and soleus (Fig. 4.2)

The exact same position is adopted for treatment as for testing, with the knee flexed over a rolled towel or cushion if soleus is being treated and the knee straight if gastrocnemius is being treated.

If the condition is acute (defined as a dysfunction/injury of less than 3 weeks' duration) the area is treated with the foot dorsiflexed to the restriction barrier.

If it is a chronic problem (longer duration than 3 weeks) the barrier is assessed and the muscle treated in a position of ease, in the mid-range, away from the restriction barrier.

Starting from the appropriate position, based on the degree of acuteness or chronicity, the patient is asked to exert a small effort (no more than 20% of available strength) towards plantar-flexion, against your unyielding resistance, with appropriate breathing (see Box 4.2 Notes on breathing, earlier in this chapter, p. 67).

This effort isometrically contracts either gastrocnemius or soleus (depending on whether the knee is unflexed or flexed). This contraction is held for 7 to 10 seconds (longer – up to 30 seconds – if condition is chronic) together with held breath (if appropriate).

On slow release, on an exhalation, the foot/ ankle is dorsiflexed (be sure to flex the whole foot and not just the toes!) to its new restriction barrier if acute, or slightly and painlessly beyond the new barrier if chronic (and if chronic, held there for 7 to 10 seconds in slight stretch).

This pattern is repeated until no further gain is achieved (backing off to mid-range for the next contraction, if chronic, and working from the new resistance barrier if acute).

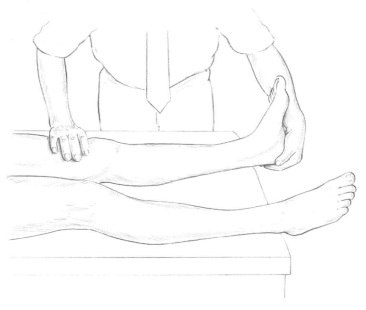

Figure 4.2 MET treatment position for gastrocnemius. If knee were flexed, the same position would focus on treatment of soleus only.

Alternatively, the antagonists to the short muscles can be used by introducing resisted dorsiflexion with the muscle at its barrier or short of it (acute/chronic) followed by <u>painless</u> stretch to the new barrier (acute) or beyond it (chronic), during an exhalation.

Use of antagonists in this way is less effective than use of agonist but may be a useful strategy if trauma has taken place.

2. Assessment and treatment of medial hamstrings and adductors

Assessing for shortness in medial hamstrings (03) (semi-membranosus and semi-tendinosus as well as gracilis) and short adductors (04) (pectineus, adductors brevis, magnus and longus)

Method (a) Patient lies so that non–tested leg is abducted slightly, heel over end of table; leg to be tested is close to edge of bed. Operator ensures that tested leg is in anatomically correct position, knee in full extension and with no external rotation of the leg, which would negate the test.

Operator should effectively stand between patient's leg and table so that all control of tested leg is achieved with lateral arm/hand and the table-side hand can rest on, and palpate, the inner thigh muscles for sensations of bind, as it is taken into abduction. Abduction of the tested leg is introduced passively until the first sign of resistance is noted (see Fig. 4.3).

Effectively, there are three indicators of this resistance:
— a sense that the motive hand carrying the leg picks up, as an increase in required effort at the moment that the first resistance barrier is passed
— the sense of bind noted by the palpating hand at this same moment
— a visual sign, movement of the pelvis as a whole, laterally towards the tested side, as the barrier is passed.

If the degree of abduction produces an angle with the mid-line of 45° or more, then no further test is needed, the abduction is normal and there is probably no shortness in the short or long adductors (medial hamstrings or, more correctly, gracilis, semi-membranosus and semi-tendinosus).

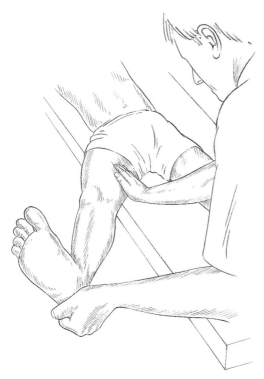

Figure 4.3 Assessment and treatment position for medial hamstrings. Short adductor shortness may be evaluated and treated in the same relative position but with the knee of the leg to be treated in flexion.

If abduction ceases before a 45° angle is easily achieved, then restriction exists in either the medial hamstrings or the short adductors of the thigh.

Screening short adductors (04) from medial hamstrings (03)

As in test number one, it is necessary to screen between shortness of the one and two joint muscles (short adductors and medial hamstrings).

This is achieved by abducting the leg to its easy barrier and then introducing flexion of the knee, allowing the lower leg to hang down freely.

If further abduction is now easily achieved to 45° (after knee flexion has been introduced) this indicates that previous limitation into abduction was the result of medial hamstring shortness,

since this is no longer operating once the knee has been flexed.

If restriction remains, or fails to release sufficiently to allow a 45° excursion when knee flexion is introduced to the abducted leg (as evidenced by movement of the pelvis or increase in palpated sense of bind in the adductors), then it is apparent that the short adductors are continuing to prevent movement, and are short.

Assessing for shortness in medial hamstring (03)/short adductors (04)

Method (b) Patient lies at very end of table (coccyx close to edge), non-tested leg fully flexed at hip and knee, and held to chest by patient (or sole of patient's foot resting against operator's lateral chest wall) to stabilise pelvis in full rotation, so that lumbar spine is not in extension.

Tested leg is grasped both above and below the knee (operator has two free hands in this position, one of which can usefully palpate the thigh for bind during the assessment) and taken into abduction to the first sign of resistance. If this reaches 45° then the test shows no shortness.

If a restriction/resistance barrier is noted before 45° then the knee should be flexed to screen the short adductors from the medial hamstrings as in method (a). In all other ways the findings are interpreted as above.

MET treatment of shortness in short and long adductors of the thigh

Precisely the same positions may be adopted for treatment as for testing, whether test method (a) or test method (b) was used.

If short adductors (pectineus, adductors brevis, magnus and longus) are being treated then the leg with knee flexed is held at the barrier (acute) or in the mid-range (chronic) while an isometric contraction is introduced using the agonists (push is away from the barrier of resistance) or the antagonists (push is towards the barrier of resistance) for 7 to 10 seconds (with appropriate breathing, see notes on breathing earlier in this chapter, Box 4.2, p. 67), using around 20% of available strength (longer and somewhat stronger for chronic than acute) followed by an easing of the leg to its new barrier (acute) or painlessly beyond and into stretch (chronic – in which case it is held there for as long as the contraction, in order to stretch fibrous and shortened tissue). Repeat until no further gain is apparent.

If medial hamstrings (semi-membranosus, semi-tendinosus as well as gracilis) are being treated, all elements are the same except that the leg is held in extension, with no bend of the knee.

Whichever position is used, the subsequent stretch, on an exhalation, is used to (acute), or through (chronic), the barrier to commence normalisation of the short muscles.

CAUTION: The major error made in treating these particular muscles relates to allowing a pivoting of the pelvis and a low spinal sidebend to occur. Maintenance of the pelvis in a stable position is vital, and this can most easily be achieved via suitable straps or, during treatment, by having the patient side-lying with the affected side uppermost, operator standing behind and using the caudad arm and hand to control the leg, flexed or straight as appropriate, while the cephelad hand maintains a firm downwards pressure on the lateral pelvis to ensure stability during stretching.

3. Assessment and treatment of hip flexors – rectus femoris, iliopsoas

Assessment of shortness in hip flexors – rectus femoris (05), iliopsoas (06) (see Fig. 4.4)

Patient lies supine with buttocks (coccyx) as close to end of table as possible, non–tested leg in flexion at hip and knee, held by patient or by having sole of foot of non-tested side placed against the lateral chest wall of the operator. This helps to maintain the pelvis in full flexion and the lumbar spine flat, which is essential if

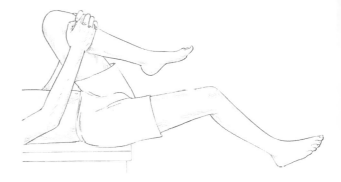

Figure 4.4B In the test position, if the thigh is elevated (i.e. not parallel with the table) probable psoas shortness is indicated. The inability of the lower leg to hang more or less vertically towards the floor indicates probable rectus femoris shortness (TFL shortness can produce a similar effect).

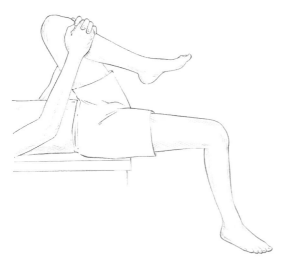

Figure 4.4A Test position for shortness of hip flexors. Note that the hip on the non-tested side must be fully flexed to produce full pelvic rotation. The position shown is normal.

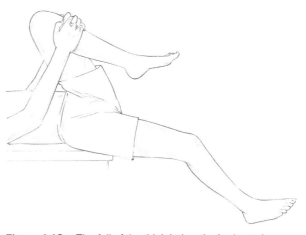

Figure 4.4C The fall of the thigh below the horizontal indicates hypotonic psoas status. Rectus femoris is once again seen to be short, while the relative external rotation of the lower leg (see angle of foot) hints at probable shortened TFL involvement.

the test is to be meaningful and stress on the spine is to be avoided.

If the thigh of the tested leg fails to rest in a horizontal position in which it is parallel to the floor/table, then the indication is that iliopsoas is short.

If the lower leg of the tested side fails to achieve an almost 90° angle with the thigh, vertical to the floor, then shortness of the rectus femoris muscle is indicated.

If this is not clearly noted, application of light pressure towards the floor on the lower third of the thigh will produce a compensatory extension of the lower leg only when rectus femoris is short.

A slight degree (10° to 15°) of hip extension should be possible in this position, by pushing downwards on the thigh, without knee extension occurring. This can subsequently be checked by seeing whether or not the heel on that side can easily flex to touch the buttock of the prone patient (if rectus is short heel will not easily reach buttock).

If the small degree of hip extension is not easily possible this confirms iliopsoas shortening.

If both psoas and rectus are short, rectus should be treated first.

If the thigh hangs down below a parallel position it indicates a degree of laxity in iliopsoas.

A further cause of failure of the thigh to rest parallel to the floor can be caused by shortness of tensor fascia lata. If this structure is short (a further test proves it, see below, p. 79) then there should be an obvious groove apparent on the lateral thigh and the patella and sometimes the whole lower leg will deviate laterally.

A further indication of short psoas is seen if the prone patient's hip is observed to remain in flexion. In this position passive flexion of the knee will result in compensatory lumbar lordosis and increased hip flexion if rectus femoris is also short.

Notes on psoas

- Lewit (1985b) mentions that in many ways the psoas behaves as if it were an internal organ. Tension in the psoas may be secondary to kidney disease, and one of its frequent clinical manifestations when in spasm, is that it reproduces the pain of gall-bladder disease (often after the organ has been removed).

- The definitive signs of psoas problems are not difficult to note, according to Harrison Fryette (1954). He maintains that the distortions produced in inflammation and/or spasm in the psoas are characteristic and cannot be produced by other dysfunction. The origin of the psoas is from 12th thoracic to (and including) the 4th lumbar, but not the 5th lumbar. The insertion is into the lesser trochanter of the femur, and thus, when psoas spasm exists unilaterally, the patient is drawn forwards and sidebent to the involved side. The ilium on the side will rotate backwards on the sacrum, and the thigh will be averted. When both muscles are involved the patient is drawn forward, with the lumbar curve locked in flexion. This is the characteristic reversed lumbar spine. Chronic bilateral psoas contraction creates either a reversed lumbar curve if the erector spinae of the low back are weak, or an increased lordosis if they are hypertonic.

- Lewit says, 'Psoas spasm causes abdominal pain, flexion of the hip and typical antalgesic (stooped) posture. Problems in psoas can profoundly influence thoraco-lumbar stability.'

- The 5th lumbar is not involved directly with psoas, but great mechanical stress is placed upon it when the other lumbar vertebrae are fixed in either a kyphotic or increased lordotic state. In unilateral psoas spasms, a rotary stress is noted at the level of 5th lumbar. The main mechanical involvement is, however, usually at the lumbodorsal junction. Attempts to treat the resulting pain, which is frequently located in the region of the 5th lumbar and sacroiliac, by attention to these areas, will be of little use. Attention to the muscular component should be a primary focus, ideally using MET.

- There exists in all muscles a vital reciprocal agonist–antagonist relationship which is of primary importance in determining their tone and healthy function. Psoas–rectus abdominus have such a relationship and this has important postural implications (see notes on 'Lower crossed syndrome' in Chapter 2, p. 33).

• Observation of the abdomen 'falling back' rather than mounding when the patient flexes, indicates normal psoas function. Similarly, if the patient, when lying supine, flexes knees and 'drags' the heels towards the buttocks (keeping them together) the abdomen should remain flat or 'fall back'. If the abdomen mounds or the small of the back arches, psoas is incompetent.

• If the supine patient raises both legs into the air and the belly mounds it shows that the recti and psoas are out of balance. Psoas should be able to raise the legs to at least 30° without any help from the abdominal muscles.

• Psoas fibres merge with (become 'consolidated' with) the diaphragm and it therefore influences respiratory function directly (as does quadratus lumborum).

• Basmajian informs us that the psoas is the most important of all postural muscles. If it is hypertonic and the abdominals are weak and exercise is prescribed to tone these weak abdominals, such as curl-ups with the dorsum of the foot stabilised, then a disastrous negative effect will ensue in which, far from toning the abdominals, increase of tone in psoas will result, due to the sequence created by the dorsum of the foot being used as a point of support. When this occurs (dorsiflexion), the gait cycle is mimicked and there is a sequence of activation of tibialis anticus, rectus femoris and psoas. If, on the other hand, the feet could be plantarflexed during curl-up exercises, then the opposite chain is activated (triceps surae, hamstrings and gluteals) inhibiting psoas and allowing toning of the abdominals.

• When treating, it is sometimes useful to assess changes in psoas length by periodic comparison of apparent arm length. Patient lies supine, arms extended above head, palms together so that length can be compared. A shortness will commonly be observed in the arm on the side of the shortened psoas, and this should normalise after successful treatment (there may of course be other reasons for apparent difference in arm length, and this method provides an indication only of changes in psoas length).

Mitchell's strength test

Before using MET methods to normalise a short psoas, Mitchell recommends that you have the patient at the end of the table, both legs hanging down and feet turned in so that they can rest on your lateral calf areas as you stand facing the patient.

The patient should press firmly against your calves with her feet as you rest your hands on her thighs and she attempts to lift you from the floor. In this way you assess the relative strength of one leg's effort as against the other. Judge which psoas is weaker or stronger than the other. If a psoas has tested short (as above) and strong then it is suitable for MET treatment according to Mitchell. If it tests short and weak then other factors such as tight erector spinae muscles should be treated first until it tests strong and short, at which time MET should be applied.

Before treating a shortened psoas, any shortness in rectus femoris on that side should first be treated.

MET treatment for shortness of rectus femoris

Patient lies prone, ideally with a cushion under the abdomen to help avoid hyperlordosis. The operator stands on the side of the table of the affected leg so that he can stabilise the patient's pelvis (hand over sacral area) during the treatment, using cephalad hand.

The affected leg, flexed at hip and knee, is cradled, so that the operator's hand curls under the lower thigh, just above the knee, as the foot rests anterior to the operator's shoulder. If rectus is short then the patient's heel will not easily be able to be taken to touch the buttock (Fig. 4.5).

Once the restriction barrier has been established (how close can the heel get to the buttock without force?) the decision is made as to

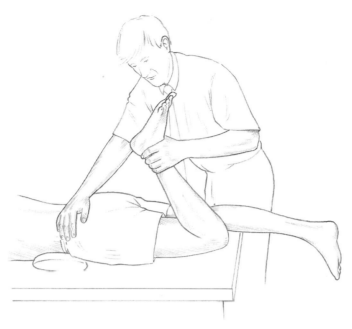

Figure 4.5   MET treatment of left rectus femoris muscle. Note the operator's right hand stabilises the sacrum and pelvis to prevent undue stress during the stretching phase of the treatment.

whether to treat as an acute problem, from the barrier, or as a chronic problem, in the mid-range.

Appropriate degrees of resisted isometric effort are then introduced (mild effort for acute, and longer, stronger efforts for chronic) in which the patient tries to both straighten the leg and take the thigh to the table (involving both ends of rectus). Use appropriate breathing (see notes on breathing earlier in this chapter, Box 4.2, p. 67). The contraction is followed, on an exhalation, by stretching of the muscle to or through the new barrier subsequently by taking the heel towards the buttock (acute and chronic modes of treatment, see previous notes, p. 66).

Remember to increase slight hip extension after each effort (perhaps using a cushion to support the thigh away from the table) as this stretches the cephalad end of rectus.

Repeat until no further gain can be achieved and then utilise antagonists (have the patient try to flex the heel towards the buttock instead of trying to straighten the leg) if this is appropriate (see MET instruction notes, p. 66).

Once a reasonable degree of increased range has been gained in rectus femoris it is appropriate to treat psoas, if this tested as short.

MET treatment of psoas (Fig. 4.6)

Psoas can be treated in the prone position described for rectus above, in which case the stretch following the patient's isometric effort to bring the thigh to the table against resistance would be concentrated on extension of the thigh, either to the new barrier of resistance if acute or past the barrier, placing stretch on psoas, if chronic.

A better position for treatment is the supine test position, in which the patient lies with the buttocks at the very end of the table, non-treated leg fully flexed at hip and knee and either held in that state by the patient or by placement of the patient's foot against the operator's lateral chest wall. The leg on the affected side is allowed to hang over the edge, with the medio-plantar aspect resting on the operator's far knee

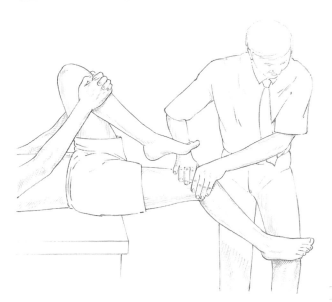

Figure 4.6A MET treatment of psoas using Grieve's method, in which there is placement of the patient's foot, inverted, against the operator's thigh. This allows a more precise focus of contraction into psoas when the hip is flexed against resistance.

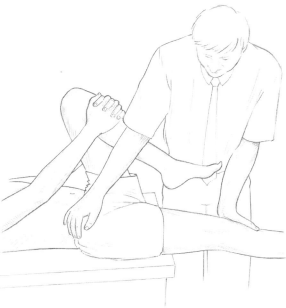

Figure 4.6B Psoas treatment variation, with the leg held straight and the pelvis stabilised.

or shin. The operator stands sideways on to the patient, at the foot of the table, with both hands holding the thigh of the extended leg. The operator's far leg should be flexed slightly at the knee, so that the patient's foot can rest, as described. This is used as a contact which, with the hands, resists the attempt of the patient to externally rotate the leg and, at the same time, flex the hip. The operator resists both efforts, and an isometric contraction of the psoas and associated muscles therefore takes place. This combination of forces focuses the contraction effort into psoas very precisely. Introduce appropriate breathing instructions if possible (see notes on breathing, Box 4.2, p. 67).

If the condition is acute the treatment of the patient's leg commences from the restriction barrier, whereas, if the condition is chronic, the leg is elevated into a somewhat more flexed position in the mid-range.

After the isometric contraction, using effort suitable to the degree of acuteness/chronicity, the thigh should, on an exhalation, either be taken to the new restriction barrier, without force (acute) or through that barrier with slight, painless force (chronic) – and held there for as long as the contraction duration. Repeat until no further gain is achieved (see Fig. 4.7).[3]

Self-treatment of psoas

Lewit suggests self-treatment in a position as above in which the patient, having taken the extended leg to the limit of its stretch, is told to lift it slightly (say 2 cm) and to breathe in slowly, and then to slowly let the knee drop as he exhales. This is repeated 3 to 5 times. The counterpressure in this effort is achieved by gravity.

4. Assessment and treatment of hamstrings

Assessment for shortness in hamstrings (07) (Fig. 4.8)

If the hip flexors (psoas etc.) were tested as short then the test position for hamstrings needs to

---

[3]Direct inhibitory pressure techniques onto the origin of psoas through the mid-line is an effective alternative approach.

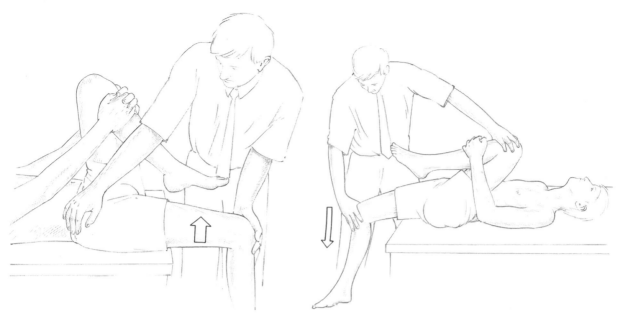

Figure 4.7A MET treatment involves the patient's effort to flex the hip against resistance.

Figure 4.7B Stretch of psoas, which follows the isometric contraction (4.7A) and is achieved by means of gravity plus additional operator effort.

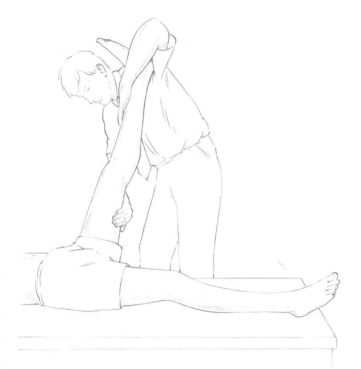

Figure 4.8A Assessment for shortness in hamstring muscles. The operator's right hand palpates for bind/the first sign of resistance, while the left hand maintains the patient's knee in extension.

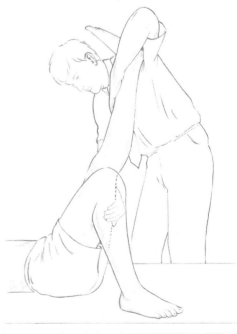

Figure 4.8B The patient's non-treated leg is flexed in this variation of the MET assessment and/or treatment of shortened hamstring muscles, in cases in which a previous assessment has indicated shortness of the hip flexors (see text page 76).

commence with the non-tested side leg flexed at knee and hip, foot resting flat on the treatment surface to ensure full pelvic rotation into neutral. If no hip flexor shortness was observed then the non–tested leg should lie straight on the surface of the table, while the operator resists its tendency to flex at the knee during straight leg raising of the tested leg, by maintaining a down-wards pressure on it just above the knee.

Test (a) Patient lies supine with non-tested leg either flexed or straight, depending on previous test results for hip flexors. The tested leg is taken into a straight leg raised position, no flexion of the knee being allowed, and minimal force employed with the barrier of restriction being assessed at the first sign of resistance.

If straight leg raising to 80° is not easily possible then there exists some shortening of the hamstrings and the muscles can be treated with the leg straight (see below, p. 79).

Test (b) Whether or not an 80° elevation is easily achieved, a variation in testing is also needed to evaluate the lower fibres.

To make this assessment the tested leg is taken into full hip flexion and the knee is then straightened.

If it cannot straighten at the knee with hip flexed this indicates shortness in the lower hamstring fibres and the patient will report a degree of pull behind the knee and lower thigh. Treatment of this is carried out in the test position.

If the knee is capable of being straightened with hip flexed, having previously not been capable of achieving a 90°, straight leg position, then the lower fibres are cleared of shortness and it is the upper fibres of hamstrings which require attention using MET, using the test position as the basis for this.

Test (c) Lewit describes a functional test which helps to screen for overactivity in the erector spinae and/or hamstrings, indicating also weak-ness of gluteus maximus.

The patient is prone and the operator places palpating hands on the lower buttock/upper thigh (gluteal/hamstring contact) and the low back (erector spinae contact) as the patient is asked to extend the hip, leg straight. The normal sequence is for the hamstrings to commence the elevation of the thigh, with almost instant gluteal involvement, followed by the erectors. If gluteus maximus is weak (see Lower Crossed Syndrome notes in Chapter 2, p. 33) there may still be strong extension of the thigh, but with the hamstrings and erectors doing most of the work. This is proof of overactivity (i.e. stress) in these postural muscles and therefore indicates shortness as being likely. In extreme cases the movement of thigh/hip extension is initiated by the erector spinae themselves, and these are then almost certainly short.[4]

MET for shortness of lower hamstrings

If the lower hamstring fibres are implicated as being short then the treatment position is identical to the test position. This means that the non-treated leg needs to be either flexed or straight on the table, depending upon whether hip flexors have previously been shown to be short or not (see above p. 72), and the treated leg needs to be flexed at the hip and the knee and then straightened until the restriction barrier is assessed (one hand may usefully palpate the tissues behind the knee for sensations of bind).

Depending upon whether it is an acute situation or a chronic one, the isometric contraction against resistance is introduced at the barrier (acute) or in the mid-range (chronic).

It is particularly important with the hamstrings to take care regarding cramp and so it is suggested that no more than 25% of patient's effort should ever be used at any time during isometric contractions.

Following the 7 to 10 seconds of contraction (holding breath if possible, see Box 4.2, Notes on

---

[4]Extension of the hip is a normal part of the gait cycle and therefore a movement which is made thousands of times daily. It is not hard to imagine the degree of overuse stress (and therefore ultimately shortness) in those postural muscles which are compensating for the weakness in the gluteal muscles (a situation for which they may well be partly responsible). If the pattern as described exists the hamstrings can be assumed to be short and out of balance with their synergists and likely to benefit from MET.

breathing, p. 67) and complete relaxation, the leg should, on an exhalation, be straightened at the knee to its new barrier (in acute problems) and through the barrier with a degree of passive stretch (if chronic). Repeat until no further gain is possible.

Antagonist muscles can also be used isometrically by having the patient try to extend the knee during the contraction rather than bending it, followed by the same stretch as would be adopted if the agonist (affected muscle) had been employed.

MET for shortness of upper hamstrings

If the upper fibres are involved then treatment is in the straight leg raising position, knee maintained in extension at all times, and the other leg flexed at hip and knee or straight, depending on the hip flexor findings as explained above (p. 77). In all particulars, the procedures are the same as for lower fibres except that the leg is kept straight.

Alternative methods

An alternative method is for the supine patient to flex the affected hip fully. The flexed knee is extended as far as possible without strain, with the back of the lower leg resting on the shoulder of the operator, who stands facing the head of the table. If this involves the right leg of the patient, then the operator's left hand will rest on the anterior aspect of patient's shin. The operator's right hand stabilises the patient's extended unaffected leg against the couch. The patient is asked to attempt to straighten the lower leg (i.e. extend the knee) utilising the antagonists to the hamstrings which require stretching, employing 20% of the strength in the quadriceps. This is resisted by the operator for 7 to 10 seconds. Introduce appropriate breathing instructions if possible (see notes on breathing, Box 4.2, p. 67). The leg is then extended at the knee to its new hamstring limit (or stretched slightly) after relaxation and the procedure is repeated.

Or, in this same position, the patient attempts to flex the knee (causing downward pressure against the operator's shoulder with the back of

the lower leg) thus employing the hamstrings themselves isometrically for 7 to 10 seconds (with appropriate breathing, see Box 4.2, p. 67). After relaxation, the muscles are taken further to or through their new barrier (acute/chronic).

Or, starting from the same position, in order to produce a combined contraction, there could be an instruction to the patient to pull the thigh towards the face (flex the hip) while the flexed lower leg was being pushed downward through the shoulder. This effectively contracts both the quadriceps and the hamstrings, thus inducing both postisometric relaxation and reciprocal inhibition, which facilitates subsequent stretching or easing to the barrier of previously tight hamstrings, on an exhalation.

5. Assessment and treatment of tensor fascia lata

Assessment of shortness in tensor fascia lata (TFL)(08)

The test recommended is a modified form of Ober's test (see Fig. 4.9).

Patient is side-lying with back close to edge of the table. Operator stands behind patient, whose lower leg is flexed at hip and knee and held in this position by the patient for stability. The

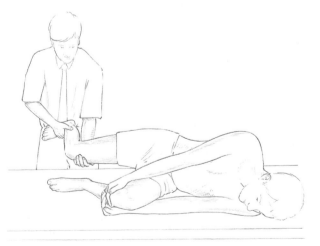

Figure 4.9 Assessment for shortness of TFL – modified Ober's test. When the hand supporting the flexed knees is removed the thigh should fall to the table if TFL is not short.

tested leg is supported by the operator, who must ensure that no hip flexion occurs which would nullify the test. The leg is abducted and extended only to the point where the iliotibial band lies over the greater trochanter. The tested leg is held by operator at ankle and knee, with the whole leg in its anatomical position, neither abducted nor adducted and not forward or backward of the body.

The operator carefully introduces flexion at the knee to 90° without allowing the hip to flex and then, holding just the ankle, allows the knee to fall towards the table.

If TFL is normal, the thigh and knee will fall easily with the knee contacting the table surface. If the upper leg remains aloft, with little sign of 'falling' then either the patient is not letting go or the TFL is tight and does not allow it to fall. The band will palpate as tender under such conditions, as a rule.

Notes on TFL

- Mennell says that TFL shortness can produce all the symptoms of acute and chronic sacroiliac problems.
- Pain from TFL shortness can be localised to the posterior superior iliac spine (PSIS), radiating to the groin or down any aspect of the thigh to the knee.
- Although the pain may arise in the sacroiliac (SI) joint, dysfunction in the joint may be caused and maintained by taut TFL structures.
- Pain from the band itself can be felt in the lateral thigh, with referral to hip or knee.
- TFL can be 'riddled' with sensitive fibrotic deposits and trigger point activity.
- There is commonly a posteriority of the ilium associated with short TFL.
- TFL's prime phasic activity (all postural structures also have some phasic function) is to assist the gluteals in abduction of the thigh.
- If TFL and psoas are short they may, according to Janda, 'dominate' the gluteals on abduction of the thigh, so that a degree of lateral rotation and flexion of the hip will be produced, rotating the pelvis backwards.

- Rolf points out that persistent exercise such as cycling will shorten and toughen the fascial iliotibial band 'until it becomes reminiscent of a steel cable'. This band crosses both hip and knee, and spatial compression allows it to squeeze and compress cartilaginous elements such as the menisci. Ultimately, it will no longer be able to compress, and rotational displacement at knee and hip will take place.

Lewit's TFL palpation

Patient is side-lying and operator stands facing the patient's front, at hip level. The operator's cephalad hand rests over the ASIS so that it can also palpate over the trochanter. It should be placed so that the fingers rest on the TFL and trochanter and the thumb on gluteus medius.

The caudad hand rests on the mid-thigh to apply slight resistance to the patient's effort to abduct the leg. The patient's table-side leg is slightly flexed to provide stability, and there should be a vertical line to the table between one anterior superior iliac spine (ASIS) and the other.

The patient abducts the upper leg (which should be extended at the knee and slightly hyperextended at the hip) and the operator should feel the trochanter 'slip away' as this is done. If, however, the whole pelvis is felt to move rather than just the trochanter, there is inappropriate muscular imbalance. In balanced abduction gluteus comes into action at the beginning of the movement, with TFL operating later in the pure abduction of the leg.

If there is an overactivity (and therefore shortness) of TFL, then there will be pelvic movement on the abduction, and TFL will be felt to come into play before gluteus. This confirms a stressed postural structure (TFL), which implies shortness.

It is possible to increase the number of palpation elements involved by having the cephalad hand also palpate (with an extended small finger) quadratus lumborum, during leg abduction.

In a balanced muscular effort to lift the leg sideways, quadratus should not become active until the leg has been abducted to around 25° to 30°. When it is overactive it will often start the abduction along with TFL, thus producing a pelvic tilt.[5]

Method (a) Supine MET treatment of shortened TFL (Fig. 4.10) Patient lies supine with unaffected leg flexed at hip and knee. Affected side leg is adducted to its barrier which necessitates it being brought under the opposite leg/foot. Using guidelines for acute and chronic problems, the structure will either be treated at or short of the barrier of resistance, using light or fairly strong isometric contractions for short (7 second) or long (up to 30 seconds) durations, using appropriate breathing patterns as described earlier in this chapter (Box 4.2, p. 67). The operator uses his trunk to stabilise the pelvis by leaning against the flexed (non-affected side) knee.

The operator's cephalad arm lies along the lateral aspect of the affected leg so that the knee is stabilised by the hand; the other hand holds the affected leg at the ankle.

After the contraction (patient abducts leg against resistance) ceases and the patient has relaxed using appropriate breathing patterns, the leg is taken to or through the new restriction barrier (into adduction past the barrier) to stretch the muscular fibres of TFL (the upper third of the structure). Care should be taken to ensure that the pelvis is not tilted during the stretch. Stability is achieved by the operator increasing pressure against flexed knee/thigh.

This whole process is repeated until no further gain is possible.

Method (b) Isolytic variation If an isolytic contraction is introduced in order to stretch actively the interface between elastic and non-elastic tissues, then there is a need to stabilise the pelvis more efficiently, either by use of wide straps or another pair of hands holding the ASIS downwards towards the table during the stretch. The procedure consists of the patient

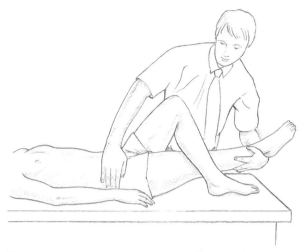

Figure 4.10 MET treatment of TFL (see Figure 1.4 on page 8 for description of isolytic variation.) If a standard MET method is being used, the stretch will follow the isometric contraction in which the patient will attempt to move the right leg to the right against sustained resistance. It is important for the operator to maintain stability of the pelvis during the procedure. Note: the hand positions in this figure are a variation of those described in the text.

attempting to abduct the leg as the operator overcomes the muscular effort, forcing the leg into adduction. The contraction/stretch should be rapid (2 to 3 seconds at most to complete). Repeat several times.

Method (c) Side-lying MET treatment of TFL Patient lies on the affected TFL side with upper leg flexed at hip and knee and resting forward of the affected leg. Operator stands behind patient and uses caudad hand and arm to raise the affected leg while stabilising pelvis with the cephalad hand, or uses both hands to raise the affected leg into slight adduction (appropriate if strapping used to hold pelvis to table). The patient contracts the muscle against resistance by trying to take the leg into abduction (towards the table) using breathing assistance as appropriate (see notes on breathing, Box 4.2, p. 67). After the effort, on an exhalation, the operator lifts the leg into adduction beyond the barrier to stretch the interface between elastic and non-elastic tissues. Repeat as appropriate or modify to use as an isolytic contraction by stretching the structure past the barrier during the contraction.

---

[5]Remember that a lateral 'corset' of muscles exists to stabilise the pelvic and low back structures and that if TFL and quadratus (and/or psoas) shorten and tighten, the gluteal muscles will weaken. This test gives the proof of such imbalance existing. (See notes on Lower Crossed Syndrome in Ch. 2, p. 33.)

Additional TFL methods

Mennell has described superb soft-tissue stretching techniques for releasing TFL. These involve a series of snapping actions applied by thumbs to the anterior fibres with patient side-lying, followed by a series of heel of hand thrusts across the long axis of the posterior TFL fibres.

Additional stretching is possible by use of elbow 'stripping' of the structure, neuromuscular deep tissue approaches to the upper fibres and those around the knee and specific deep release methods. All of these require expert tuition to allow acquisition of skills.

Self-treatment and maintenance

The patient lies on the side, on a bed or table, with the affected leg uppermost and hanging over the edge (lower leg comfortably flexed). The patient may then use postisometric relaxation by slightly lifting the hanging leg (2 cm or so) whilst inhaling, and holding this position for a matter of 10 seconds or so, before releasing and allowing a greater degree of stretch to take place. This is then repeated several times in order to achieve the maximum available stretch in the tight soft tissues. The counterforce in this isometric exercise is gravity.

6. Assessment and treatment of piriformis

Assessment of shortened piriformis (09)

Test (a) Stretch test When short, piriformis will cause the affected side leg of the supine patient to appear to be short and externally rotated.

With the patient supine, place the tested leg into flexion at hip and knee, so that the foot rests on the table lateral to the other knee (the tested leg is crossed over the straight non-tested leg, in other words, as shown in Fig. 4.11).

The non-tested side ASIS is stabilised to prevent pelvic motion during the test and the knee of the tested side is pushed into adduction to place a stretch on piriformis. If there is a short piriformis the degree of adduction will be limited and the patient will report discomfort behind the trochanter.

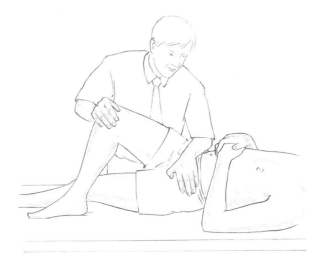

Figure 4.11 MET treatment of piriformis muscle with patient supine. The pelvis must be maintained in a stable position as the knee (right in this example) is adducted to stretch piriformis following an isometric contraction.

Test (b) Palpation test Patient is side-lying, tested side uppermost. Operator stands at the level of the pelvis in front of and facing the patient, and in order to contact the insertion of piriformis, draws imaginary lines between:

— ASIS and ischial tuberosity, and
— PSIS and the most prominent point of trochanter.

Where these lines cross, just posterior to the trochanter, is the insertion of the muscle, and pressure here will produce marked discomfort if the structure is short or irritated.

If the most common trigger point site in the belly of the muscle is sought, then the line from the ASIS should be taken to the tip of the coccyx rather than the ischial tuberosity. Pressure where this line crosses the other will access the mid-point of the belly of piriformis where triggers are common. Light compression here which produces a painful response is indicative of a stressed muscle.

Notes on piriformis

• This postural muscle, like all others, will shorten if stressed. In the case of piriformis, the effect of shortening is to increase its diameter and, because of its location, this allows for direct pressure to be exerted on the sciatic nerve, which passes under it in 80% of people. In the other 20% it passes through the muscle so that contraction will produce veritable strangulation of the nerve.

- In addition, the pudendal nerve and the blood vessels of the internal iliac artery, as well as common perineal nerves, posterior femoral cutaneous nerve and nerves of the hip rotators, can all be affected.

- If there is sciatic pain associated with piriformis shortness then on straight leg raising, which reproduces the pain, external rotation of the hip should relieve it, since this slackens piriformis.

- This clue may, however, only apply to any degree if the individual is one of those in whom the nerve actually passes through the muscle.

- The effects can be circulatory, neurological and functional, inducing pain and paraesthesia of the affected limb as well as alterations to pelvic and lumbar function. Diagnosis usually hinges on the absence of spinal causative factors and the distributions of symptoms from the sacrum to the hip joint, over the gluteal region and down to the popliteal space. Palpation of the affected piriformis tendon, near the head of the trochanter, will elicit pain. The affected leg will be externally rotated.

- The piriformis muscle syndrome is frequently characterised by such bizarre symptoms that they may seem unrelated. One characteristic complaint is a persistent, severe, radiating low back pain extending from the sacrum to the hip joint, over the gluteal region and the posterior portion of the upper leg, to the popliteal space. In the most severe cases the patient will be unable to lie or stand comfortably, and changes in position will not relieve the pain. Intense pain will occur when the patient sits or squats since this type of movement requires external rotation of the upper leg and flexion at the knee.

Compression of the pudendal nerve and blood vessels which pass through the greater sciatic foramen and re-enter the pelvis via the lesser sciatic foramen, is possible because of piriformis contracture. Any compression would result in impaired circulation to the genitalia in both sexes. Since external rotation of the hips is required for coitus by women, pain noted during this act could relate to impaired circulation induced by piriformis dysfunction. This could also be a basis for impotency in men.

- Piriformis involvement often relates to a pattern of pain which includes:
  — pain near the trochanter
  — pain in the inguinal area
  — local tenderness over the insertion behind trochanter
  — SI joint pain on the opposite side
  — externally rotated foot on the same side
  — pain unrelieved by most positions with standing and walking being the easiest
  — limitation of internal rotation of the leg which produces pain near the hip
  — short leg on the affected side.

- The pain itself will be persistent and radiating covering anywhere from the sacrum to the buttock, hip and leg including inguinal and perineal areas.

- Bourdillon suggests that piriformis syndrome and SI joint dysfunction are intimately connected and that recurrent SI problems will not stabilise until hypertonic piriformis is corrected.

- Janda points to the vast amount of pelvic organ dysfunction to which piriformis can contribute due to its relationship with circulation to the area.

- Mitchell suggests that (as in psoas example above, p. 74) piriformis shortness should only be treated if it is tested to be short and stronger than its pair.

- If it is short and weak (see below for strength test) then whatever is hypertonic and influencing it should be released and stretched first, according to Mitchell. When it tests strong and short, piriformis should receive MET treatment.

- Since piriformis is an external rotator of the hip it can be inhibited (made to test weak) if an internal rotator such as TFL is hypertonic or if its pair is hypertonic, since one piriformis will inhibit the other.

Strength test

Patient lies prone, both knees flexed to $90°$ with operator at foot of table grasping lower legs at the limit of their separation (which internally rotates the hip and therefore allows comparison of range of movement permitted by shortened external rotators such as the piriformis).

The patient attempts to bring the ankles together as the operator assesses the relative strength of two legs.

If there is relative shortness as evidenced by the lower leg not being able to travel as far from the mid-line as its pair in this position, and if that same side also tests strong, then MET is called for.

If there is shortness but also weakness then the reasons for the weakness need to be dealt with prior to stretching using MET.

MET treatment of piriformis

Method (a) Side-lying Patient lies on side, close to edge of table, affected side uppermost, both legs flexed at hip and knee. Operator stands facing patient at hip level. Operator places cephalad elbow tip gently on access point behind trochanter, where piriformis inserts. Patient is close enough to edge of table for operator to stabilise the pelvis against his trunk (Fig. 4.12).

At same time, operator's caudad hand grasps ankle and brings upper leg/hip into internal rotation, stretching piriformis.

A degree of inhibitory pressure is applied via the elbow for 5 to 10 seconds while the muscle is kept at a reasonable but not excessive degree of stretch. The operator maintains contact but eases pressure and asks the patient to introduce an isometric contraction in piriformis by bringing the lower leg towards the table against resistance. Use the same acute and chronic rules as discussed previously (see p. 66) as well as suitable degrees of effort and duration, and

Figure 4.12 A combined ischaemic compression (elbow pressure) and MET side-lying treatment of piriformis. The pressure is alternated with isometric contractions/stretching of the muscle until no further gain is achieved.

cooperative breathing if appropriate (see Box 4.2, p. 67). After the contraction ceases and the patient relaxes, reapply elbow pressure after taking the lower limb to its new resistance barrier, allowed by release of piriformis tension. Repeat until no further gain is achieved.

This is a variation on the method advocated by TePoorten (1960) which calls for longer and heavier compression and no intermediate isometric contractions.

Method (b)i In the first stage of TePoorten's method the patient lies on the non-affected side with knees flexed and the upper legs perpendicular to the body. The operator places his elbow on the piriformis musculotendinous junction and a steady pressure of 20–30 lbs (9–13 kg) is applied. With his other hand he abducts the foot so that it will force an internal rotation of the upper leg. The leg is held in the rotated position for periods of up to 2 minutes. This procedure is repeated 2 or 3 times. The patient is then placed in the supine position and the affected leg is tested for freedom of both external and internal rotation.

Method (b)ii The second stage of TePoorten's treatment is performed with the patient in the supine position with both legs extended. The foot of the affected leg is grasped and the leg is flexed at both the knee and the hip. At the same time the foot is turned inward, which forces an external rotation of the upper leg. When the operator causes the patient's leg to be extended, the foot is turned outward, resulting in an internal rotation of the upper leg. During the procedure the patient is instructed to resist the movements (isokinetic activity); this accomplishes a modified form of isometric activity of the muscles. This treatment method, repeated 2 or 3 times, also serves to relieve the contracture of the muscles of external and internal rotation.

Method (c) A series of MET isometric contractions and stretches can be applied with the patient prone and the affected side knee flexed, the lower leg being used as a lever as the leg is taken laterally, rotating the hip internally and putting piriformis at stretch. (Use acute and chronic guidelines, p. 66, to determine the appropriate starting point for the contraction.) The patient's isometric contraction involves bringing the heel back towards the midline against resistance, followed by a move to or through the barrier as appropriate.

Application of inhibitory pressure in this position is possible via thumb but is not necessary.

Method (d) Lewit has the patient supine, the affected leg crossed over the straight unaffected leg. Adduction pressure is applied to the affected side knee to stretch piriformis, and the treatment commences at this barrier if acute, or short of it if chronic, with the usual sequence of contraction/relaxation/stretch to new barrier or through it, whichever is more appropriate, with breathing cooperation.

Method (e) A general approach which balances muscles of the region as well as pelvic diaphragm, is achieved by having patient squat while operator stands and stabilises both shoulders, preventing the patient from rising as they try to do so, while holding their breath in. After 7 to 10 seconds the effort is released; a deeper squat is performed, and the procedure is repeated several times.

Method (f) Since one piriformis contracting inhibits its pair, it is possible to self-treat an affected short piriformis by having the patient lie up against a wall with the non-affected side touching it, both knees flexed (modified from: Retzlaff et al 1974). The patient monitors the affected side by palpating behind the trochanter ensuring that no contraction takes place there.

After a contraction lasting 10 seconds or so (knee against wall) of the non-affected side, the position described in (d) above is adopted and the patient pushes the affected side knee into adduction, stretching piriformis on that side. Repeat several times.

Method (g) Stiles' piriformis technique (Stiles 1984) is as follows:

With the patient supine, and the knee on the painful side flexed to 60–70 degrees, face the patient and have him, or her, rest the back of the calf, just above the ankle, on your shoulder. Clasp your hands over the knee to stabilise it, Say to the patient: 'Gently push your heel down on my shoulder, and pull your knee towards you.' Have him/her push down and pull the knee against your resistance for a few seconds, then relax. The knee should then be extended to its new barrier, by pulling the knee caudad towards you. Repeat the sequence 2 or 3 times. This manoeuvre involves contraction of the quadriceps, the muscle group opposed to piriformis, and reciprocal innervation helps relax the muscle [if it is] in spasm. Since it also uses muscles that flex the knee (hamstring and posterior muscles of the hip) it inhibits these muscles directly (postisometric relaxation) by stimulating their Golgi receptors. In general these procedures can improve results of straight leg raising tests, when the poor results are due to hamstring or piriformis spasm.

Box 4.4   Working and resting muscles

Richard (1978) reminds us that a working muscle will mobilise up to 10 times the quantity of blood mobilised by a resting muscle. He points out the link between pelvic circulation and lumbar, ischiatic and gluteal arteries and the chance this allows to engineer the involvement of 2400 square metres of capillaries by using repetitive pumping of these muscles (including piriformis).

The therapeutic use of this knowledge involves the patient being asked to repetitively contract both piriformis muscles against resistance. The patient is supine knees bent, feet on the table; the operator resists their effort to abduct their flexed knees, using pulsed muscle energy approach (Ruddy's method) in which two isometrically resisted pulsations/contractions per second are introduced for as long as possible (a minute seems a long time doing this).

7. Assessment and treatment of quadratus lumborum

Assessment of shortness in quadratus lumborum (10) (Fig. 4.13)

Review Lewit's functional palpation test described under the heading for testing for shortness of tensor fascia lata (p. 80).

When the leg of the side-lying patient is abducted and the operator's palpating hand senses that quadratus becomes involved in this process before the leg has reached at least 25° of elevation, then it is clear that quadratus is overactive. If it is overactive then it is almost certainly stressed, and since a postural muscle which is stressed will always become short, shortness and a need for MET can be assumed.

Test (a) With the patient side-lying, have them take their upper arm over their head to grasp the top edge of the table. This 'opens out' the lumbar area and with the operator standing and facing the front of the patient allows easy palpation of the quadratus lateral border – a major trigger point site – with the cephalad hand. Test for activity of quadratus on leg abduction with the cephelad hand while also palpating gluteus medius with the caudad hand. If they act simultaneously, or if quadratus comes in first, then it is stressed and needs stretching.

Test (b) Have the patient stand, back towards you, equalise any leg length disparity by using a thin book or pad under the short leg side foot and, with the patient's feet shoulder width apart, introduce pure sidebending, running their hand down the lateral thigh/calf. Normal level of sidebending excursion is to just below the knee. Judge to which side they travel furthest. If sidebending to one side is limited then quadratus on the opposite side is probably short. Combined evidence from palpation and sidebending test indicate whether it is necessary to treat quadratus or not.

A further observation from Janda is that, in this sidebending position, 'when the lumbar spine appears straight, with compensatory motion occurring only from the thoraco-lumbar region upwards, tightness of quadratus lumborum may be suspected.'

This 'whole lumbar spine' involvement differs from a segmental restriction which would probably involve only a part of the lumbar spine.

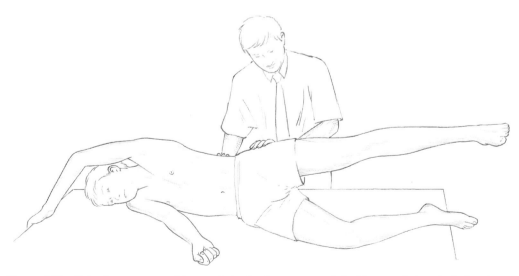

Figure 4.13   Palpation assessment for quadratus lumborum overactivity. The muscle is palpated, as is gluteus medius, during abduction of the leg. The correct firing sequence should be gluteus, followed at around 25° elevation by quadratus. If there is an immediate 'grabbing' action by quadratus it indicates overactivity, and therefore stress, so shortness can be assumed (see details of similar functional assessments in Chapter 5).

Notes on quadratus lumborum

- Quadratus fibres merge with the diaphragm (as do those of psoas) which makes involvement in respiratory dysfunction a possibility since it plays a role in exhalation.
- Shortness of quadratus, or the presence of trigger points, can result in pain in the lower ribs and along the iliac crest if the lateral fibres are affected.
- Shortness of the medial fibres, or the presence of trigger points, can produce pain in the sacroiliac joint and the buttock.
- Bilateral contraction produces extension and unilateral contraction produces extension and sidebending to that side.
- The important transition region, the thoracolumbar junction, is the only one in the spine in which two mobile structures meet and dysfunction results in alteration of the quality of motion between these structures (upper and lower trunk/dorsal and lumbar spines). In dysfunction there is often a degree of spasm or tightness in the muscles which stabilise the region, notably: psoas and erector spinae of the thoracolumbar region, as well as quadratus lumborum and rectus abdominus.

  Symptomatic diagnosis of muscle involvement is possible as follows:

  — psoas involvement usually triggers abdominal pain if severe and produces flexion of the hip and the typical antalgesic posture of lumbago

  — erector spinae involvement produces low back pain at its caudad end of attachment and interscapular pain at its thoracic attachment (as far up as the mid-thoracic level)

  — quadratus lumborum involvement causes lumbar pain and pain at the attachment of the iliac crest and lower ribs

  — rectus abdominus contraction may mimic abdominal pain and result in pain at the attachments at the pubic symphysis and the xyphoid process, as well as forward bending of the trunk and restricted ability to extend the spine.

  There is seldom pain at the site of the lesion in thoracolumbar dysfunction. Lewit points out that even if a number of these muscles are implicated it is seldom necessary, using PIR methods, to treat them all since, as the muscles most involved (based on tests for shortness, overactivity, sensitivity and direct palpation) are stretched and normalised, so will others begin to normalise themselves.

MET for shortness in quadratus lumborum

Method (a) Operator stands behind side-lying patient, at waist level. The patient has uppermost arm extended over the head to firmly grasp the top end of the table and, on an inhalation, abducts the uppermost leg until the operator palpates strong quadratus activity (elevation of around 30° usually).

The patient holds the leg (and the breath) isometrically in this manner, allowing gravity to provide resistance. After the 10 second (or so) contraction, the patient allows the leg to hang slightly behind him over the back of the table; the operator straddles this and, cradling the pelvis with both hands (fingers interlocked over crest of pelvis), leans back to take out all slack and to 'ease the pelvis away from the lower ribs' during an exhalation (see Fig. 4.14). The stretch should be held for some 10 seconds. The method will only be successful if the patient is holding the top edge of the table and so providing a fixed point from which the operator can induce stretch. This action (contraction followed by stretch) is repeated once or twice more with raised leg in front of, and once or twice with raised leg behind the trunk. The direction of stretch should be varied so that it is always in the same direction as the long axis of the abducted leg. This calls for the operator changing from the back to the front of the table for the best results when raised leg is in front of the trunk.

When the leg hangs to the back of the trunk the long fibres of the muscle are mainly affected; and when it hangs forward of the body the diagonal fibres are mainly affected, and therefore stretched.

Method (b) Gravity induced postisometric relaxation of quadratus lumborum – self-treatment.

The patient stands, legs apart, bending sideways. The patient inhales and slightly raises the

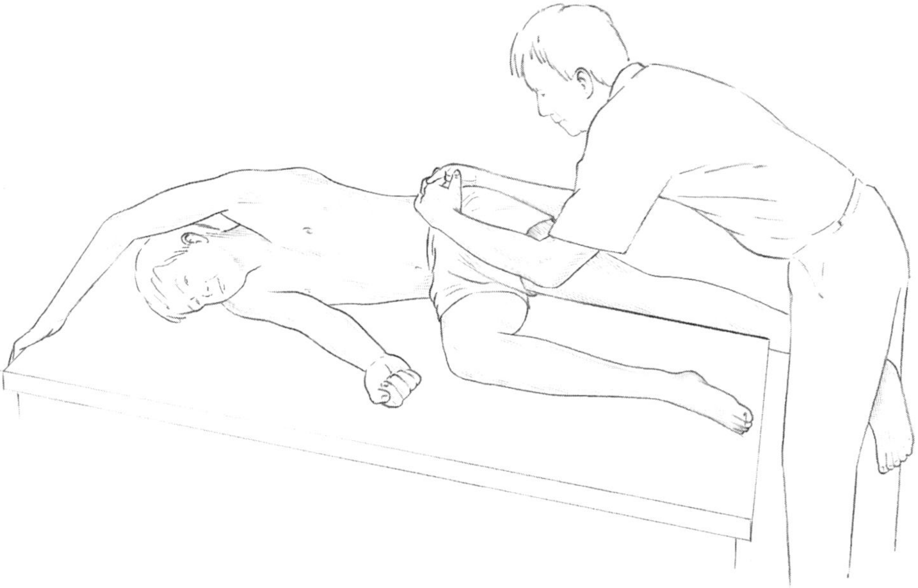

Figure 4.14 MET treatment of quadratus lumborum. Note that it is important after the Isometric contraction (sustained raised/abducted leg) that the muscle be eased into stretch, avoiding any defensive or protective resistance which sudden movement might produce. For this reason, bodyweight should be used to apply traction rather than arm strength.

trunk (a few centimetres) at the same time as looking towards the ceiling (with the eyes only). On exhalation, the sidebend is allowed to slowly go as far as it can, whilst the patient looks towards the floor. Care is needed that very little, if any, forward or backward bending is taking place at this time. This sequence is repeated a number of times (see p. 60).

Eye positions influence the tendency to flex, and sidebend (eyes look down) and extend (eyes look up). Gravity-induced stretches of this sort require holding the stretch position for at least as long as the contraction, and ideally longer. More repetitions may be needed with a large muscle such as quadratus, and home stretches should be advised several times daily.

Method (c) The side-lying treatment of latissimus dorsi described below (p. 91) also provides an effective quadratus stretch when the stabilising hand rests on the pelvic crest.

8. Assessment and treatment of pectoralis major and latissimus dorsi

Assessment of shortened pectoralis major (11) and latissimus dorsi (12)

Test (a) Observation is as accurate as most palpation for evidence of pectoralis major shortening. The patient will have a rounded shoulder posture – especially if the clavicular aspect is involved.

Test (b) The patient lies supine with the head several feet from the top edge of the table, and is asked to rest arms, extended above the head, on the treatment surface, palms upwards (Fig. 4.15).

If these muscles are normal, the arms should be able to easily reach the horizontal whilst being directly above the shoulders, in contact with the surface for almost all of the length of the upper arms, with no arching of the back or twisting of the thorax.

If an arm cannot reach the vertical above the shoulder but is held laterally, elbow pulled outwards, then latissimus dorsi is probably short on that side.

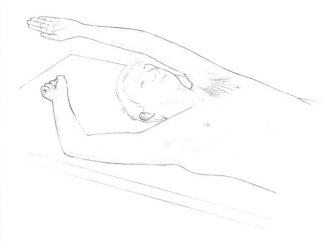

Figure 4.15   Assessment of shortness in pectoralis major and latissimus dorsi. Visual assessment is used: if the arm on the tested side is unable to rest along its full length, shortness of pectoralis major is probable; if there is obvious deviation of the elbow laterally, probable latissimus shortening is indicated.

If an arm cannot rest with the dorsum of the upper arm in contact with the table surface, without effort, then pectoral fibres are almost certainly short.

Test (c) To screen pectoralis major (11), stand at the side of the upright patient and, with the patient-side hand, stabilise the shoulder from above, fingertips resting anteriorly, close to the clavicle, holding the area (and therefore the scapula) towards the floor. With the lateral hand, cup the patient's arm (which is bent to 90°) at the elbow and abduct this slowly, taking care to avoid rotation such as would occur if the lower arm dropped taking the humerus into internal rotation. A normal degree of abduction, with no shortening of pectoralis major, or restriction in the axilla region, should allow the elbow to reach a point an inch or two from the ear, without strain.

Test (d) Assessment of the subclavicular portion of pectoralis involves abduction of the arm to 90° from the body. In this position the tendon of pectoralis at the sternum should not be found to be unduly tense, even with maximum abduction of the arm.

Test (e) To screen latissimus dorsi (12), the standing patient is asked to forward bend and

allow the arms to hang freely from the shoulders as she holds a half-bend position, trunk parallel with the floor. If the arms are hanging other than perpendicular to the floor there is probably some muscular restriction involved, and if this involves latissimus the arms will be held closer to the legs than perpendicular (if they hang markedly forward of such a position then trapezius shortening is probable, see below p. 92).

To screen latissimus in this position, one side at a time, stand in front of the patient (who remains in this half-bend position) and, stabilising the scapula area with one hand, grasp the arm at elbow level and gently draw the tested side (straight) arm forwards. It should, without undue effort or excessive 'bind' in the tissues being held, allow itself to be taken to a position where the elbow is higher than the level of the back of the head. If this is not possible, then latissimus is short.

Test (f) See assessment (b) for levator scapulae (p. 97) for a functional test which, if positive, also implicates pectoralis major (and minor) as being shortened.

MET treatment of short pectoralis major

Method (a) The patient adopts test position, so that arm is abducted in a direction which produces the most marked evidence of pectoral shortness assessed by palpation and visual evidence of the particular fibres involved.

The more elevated the arm (i.e. the closer to the head), the more abdominal the attachments will be that are being treated, and the more lateral the arm, the more clavicular the fibres. Between these two extremes lies the position which influences the sternal fibres most directly.

The patient should be as close to the side of the table as possible so that the abducted arm can be brought below the horizontal in order to apply gravitational and passive stretch to the fibres, as appropriate.

The operator stands on the side to be treated and grasps the humerus whilst the other hand contacts the insertion of the shortened fibres, attaching to a rib, or near the sternum or

Figure 4.16A   MET treatment of pectoral muscle – abdominal attachment. Note that the fibres being treated are those which lie in line with the long axis of the humerus.

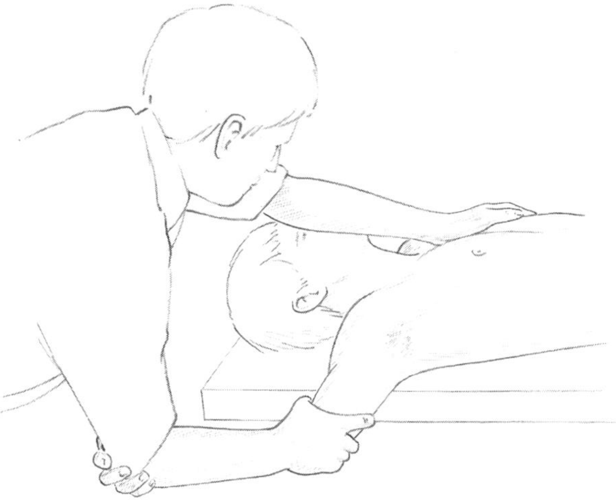

Figure 4.16B   An alternative hold for application of MET to pectoral muscle – sternal attachment. Note that the patient needs to be close to the edge of the table in order to allow the arm to be taken towards the floor once the slack has been removed, during the stretching phase after the isometric contraction.

clavicle, depending upon which fibres are being treated and therefore which arm position has been adopted (Fig. 4.16A). This contact hand stabilises the area during the contraction and stretch, preventing movement of it, but not exerting any pressure to stretch it. All stretch is via the positioning and leverage of the arm. As a rule, the long axis of the patient's upper arm should be in a straight line with the fibres being treated.

A useful hold, which depends upon the relative size of the patient and the operator, involves the operator grasping the anterior aspect of the patient's flexed upper arm just above the elbow, while the patient cups the operator's elbow and holds this contact throughout the procedure (Fig. 4.16B).

Starting with the patient's arm in a position which takes the affected fibres to or short of the restriction barrier, as appropriate (acute/chronic), the patient introduces adduction and/or elevation, against resistance, for 7 to 10 seconds with appropriate breathing (see notes on breathing, Box 4.2, p. 67).

It is important to both observe and palpate any shortened fibres (possibly housing trigger points) during the isometric contraction so that their involvement is ensured.

This is followed by a stretch to or through the new barrier, applied by the operator as the patient exhales following complete relaxation of the area. The stretch needs to be one in which the arm is pulled away from the thoracic/clavicular end as well as being taken below the horizontal if possible (distracted from the trunk and extended at the shoulder) whilst at the same time the thoracic insertion of the muscle is stabilised to prevent any movement from that point. If the stretch procedure can be thought of as having two phases (taking out the slack by distraction away from the contact/stabilising hand followed by a movement towards the floor, initiated by the operator bending the knees) there will be less danger of causing injury which might result from overenthusiastic stretching.

Method (b) Bilateral MET stretching of pectoralis major (sternocostal aspects) involves having the patient supine, knees and hips flexed, in order to provide stability to the spinal regions, preventing lumbar lordosis. A shallow but firm cushion should be placed between the scapulae, allowing a better excursion of the shoulders during this stretch. The chin should be tucked in, and if more comfortable, a small cushion placed under the neck. Ideally a strap/belt should be used to fix the thorax to the table, but this is not essential.

The operator stands at the head of the table and grasps the patient's elbows or forearms, which are flexed, laterally rotated and held in a position to induce the most taut aspects of the muscles to become prominent. Starting from such a barrier or short of it (acute/chronic) the patient is asked to contract the muscles by bringing their arms upwards and towards the table for 10 seconds or so during a held breath.

Following the contraction and complete relaxation, the arms are taken to a new or through the restriction barrier as appropriate, during an exhalation. Repeat as necessary several times more.[6]

Method (c) By adopting the same positions, but having the arms of the patient more laterally placed, so that they are laterally rotated and in 90° abduction from the shoulder (upper arms are straight out sideways from the shoulder) and there is 90° flexion at the elbows, with the operator contacting the area just proximal to the flexed elbows, a more direct stretch of the clavicular insertions of the muscle can be achieved, using all the same contraction and stretch elements as in (b) above.[6,7]

MET treatment of short latissimus dorsi

The patient is side-lying, affected side up. The arm is taken into abduction to the point of resistance, so that it is possible to visualise or palpate the insertion of the shortened fibres on the lateral chest wall. The condition is treated in either the acute or chronic mode of MET, at or short of the barrier, as appropriate.

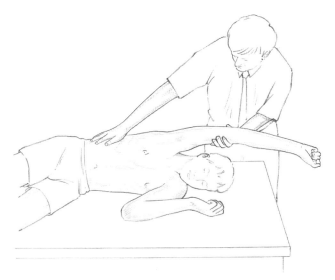

Figure 4.17 Treatment of latissimus dorsi. A variety of different positions are required for the stabilising hand (on the chest wall as well as on the crest of the pelvis) to allow for precise application of stretches of fibres with different attachments, following the sequence of isometric contractions.

As shown in Figure 4.17, the operator stands near the head of the patient, slightly behind them, and holds the upper arm in the chosen position while applying the other hand to stabilise the lateral chest wall or posterior thorax area, or even the pelvic crest, from where the stretch will be made and where tension is noted as the patient introduces an isometric contraction which attempts to bring the arm upwards (to the ceiling) and backwards and down (towards their own lower spine) against firm resistance, using only a modest amount of effort (20%) and holding the breath if appropriate (see notes on breathing, Box 4.2, p. 67).

After 10 seconds, both the effort and breath are released and the patient relaxes completely, at which time the operator introduces stretch to or through the barrier (acute/chronic), bringing the humerus into greater abduction whilst applying a stretching/stabilising contact to various points (separate contractions and stretches for each) anywhere between the lateral chest wall and the crest of the pelvis. A downward effort towards the floor assists the stretch (as in the stretch of pectoralis major, there

---

[6]Methods (b) and (c) can be adapted to a seated position, if the back is well supported (on a chair back with a cushion for example). All other elements stay the same.
[7]Serratus anterior will also be stretched by these treatment procedures. There is no specific assessment test for serratus but its shortness will be noted by sensitive attachment sites on the anterior axillary line.

should be two phases – a distraction, taking out the slack, and a movement towards the floor of the operator, by flexing the knees – to induce a safe stretch). Repeat as necessary.

Ultimately, it should be possible to achieve complete elevation of the arm without stress or obvious shortness in levator fibres so that the upper arm can rest alongside the ear of the supine patient.[8]

9. Assessment and treatment of upper trapezius

Assessment for shortness of upper trapezius (13) (Fig. 4.18)

Test (a) Lewit simplifies the need to assess for shortness by stating, 'The upper trapezius should be treated if tender and taut.' Since this is an almost universal state in modern life, it seems that everyone requires MET application to this muscle. He also notes that a characteristic mounding of the muscle can often be observed when it is very short, producing the effect of 'Gothic shoulders', similar to the architectural supports of a Gothic church tower (see Fig. 2.6, p. 34).

Test (b) The patient is supine with the neck sidebent away from, and flexed and rotated towards, the side to be tested. At this point the operator, standing at the head of the table, uses a contact on the shoulder (tested side) to assess the ease with which it can be depressed (moved distally). There should be an easy springing sensation as the shoulder is pushed towards the feet, with a soft end-feel to the movement. If there is a harsh, sudden end-point, the trapezius is probably short.

Test (c) Patient is seated and operator stands behind, one hand resting on shoulder of side to be tested. The other hand is placed on the side of the head which is being tested and the head/ neck is taken into sidebending away from that side, without force, whilst the shoulder is

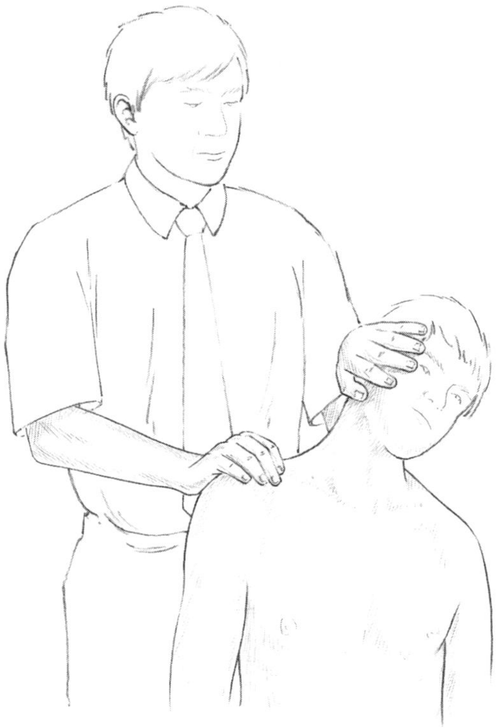

Figure 4.18 Assessment of the relative shortness of the right side upper trapezius. One side is compared with the other (for both the range of unforced motion and the nature of the end-feel of motion) to ascertain the side most in need of MET attention.

stabilised. The same procedure is performed on the other side with the opposite shoulder stabilised.

A comparison is made as to which side-bending manoeuvre produced the greater range and whether the neck can easily reach a 45° angle from the vertical, which it should. If neither side can achieve this degree of sidebend, then both trapezius muscles may be short. The relative shortness of one, compared with the other, is evaluated.[9]

Test (d) The patient is seated and the operator stands behind with a hand resting on the

---

[8]When the contact/stabilising hand is on the crest of the pelvis, the stretch using the arm as a lever will effectively also stretch quadratus lumborum.

[9]If stabilisation were made of the shoulder towards which the head is being sidebent, then assessment would be being made of the mobility of the cervical structures. By stabilising the side from which the bend is taking place, the muscular component is being evaluated.

shoulder on the side to be assessed. The patient is asked to extend the shoulder, bringing the flexed arm/elbow backwards. If the upper trapezius is very stressed/short on that side, it will inappropriately activate during this arm movement, providing cause for shortness in it to be assumed.

Test (e) See assessment (b) for levator scapulae (p. 97) for a functional test which, if positive, also implicates upper trapezius as being shortened.

MET treatment of shortened upper trapezius

There is some disagreement as to the head/neck rotation position as described in the treatment methods (a) and (b) below, which call for side-bending and rotation **away** from the affected side. Liebenson, in contrast, suggests that, 'The patient lies supine with the head supported in anteflexion and laterally flexed **away** and rotated **towards** the side of involvement'.

Lewit (see method (c) below) suggests 'The patient is supine ... the therapist fixes the shoulder from above with one hand, side-bending the head and neck with the other hand so as to take up the slack. He then asks the patient to look towards the side away from which the head is bent, resisting the patient's automatic tendency to move towards the side of the lesion.'

The author has used the methods described below with good effect and urges readers to try these and Liebenson's approaches and to evaluate results for themselves.

Method (a) The patient lies supine, head/neck sidebent away from the side to be treated, to or short of the restriction barrier (acute or chronic as appropriate) with the operator stabilising the shoulder with one hand and cupping the ear/mastoid area of the same side of the head with the other.

In order to bring into play all the various fibres of the muscle, this stretch needs to be applied with the neck in the following three different positions of rotation, coupled with the sidebending as described (see Fig. 4.19):

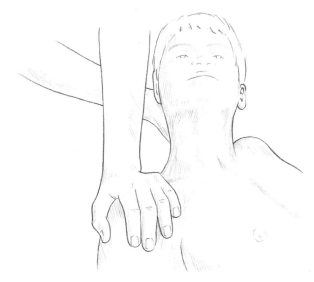

Figure 4.19 MET treatment of right side upper trapezius muscle. The head is in the upright position, with no rotation, in this example, which indicates that the anterior fibres are being treated. Note that stretching in this (or any of the alternative positions which access the middle and posterior fibres) is achieved following the isometric contraction by means of an easing of the shoulder away from the stabilised head, with no force being applied to the neck and head itself.

- With the neck sidebent and fully rotated the posterior fibres of upper trapezius are involved in any contraction, and therefore stretch.
- With the neck fully sidebent and half rotated the middle fibres are accessed.
- With the neck fully sidebent and not rotated at all the anterior fibres are being treated.

This manoeuvre can be performed with operator's arms crossed, hands stabilising the mastoid area and shoulder, or not, as comfort dictates, and with operator standing at the head or the side, also as comfort dictates.

The patient introduces a resisted effort to take the stabilised shoulder towards the ear (a shrug movement) and the ear towards the shoulder. The double movement (or effort towards movement) is important in order to introduce a contraction of the muscle from both ends.

The degree of effort should be mild and no pain should be felt. After the 10 second (or so) of contraction and complete relaxation of effort, the operator gently eases the head/neck into an increased degree of sidebending (back to the barrier if this has been reduced before the contraction in a chronic setting, or to the new barrier if it was an acute problem being treated from the resistance barrier) before stretching the shoulder away from the ear whilst stabilising the head, to or through the new barrier of resistance as appropriate. No stretch is introduced from the head-end of the muscle as this could stress the neck unduly.

Method (b) Patient is supine. Operator stands at head and places hands, arms crossed, palm downwards on anterior surface of shoulders. Patient's head rests on crossed forearms which take it and the neck into flexion until a restricted feel is noted. The patient's head is sidebent and rotated away from the side to be treated, and (either from or short of the restriction barrier depending upon degree of chronicity) a mild resisted contraction is initiated by the patient against the forearms, as the patient tries to take the head/neck back towards the table, with appropriate breathing (see Box 4.2, p. 67), for 7 to 10 seconds. After complete relaxation, and on an exhalation, the degree of flexion and sidebending/rotation is increased to or through the restriction barrier, using the forearms as levers.

This sequence is repeated with the head/neck in varying degrees of sidebending rotation so as to involve different fibres, as per method (a) above.

Method (c) Lewit suggests the use of eye movements to facilitate initiation of PIR before stretching, an ideal method for acute problems in this region. The patient is supine, while the operator fixes the shoulder and the sidebent (away from the treated side) head and neck at the restriction barrier and asks the patient to look, with the eyes only (i.e. not to turn the head) towards the side away from which the neck is bent. This eye movement is maintained as is a held breath, while the operator resists the slight isometric contraction that these two factors (eye movement and breath) will have created.

On exhalation and complete relaxation, the head/neck is taken to a new barrier and the process repeated. If the shoulder is brought into the equation this is firmly held as it attempts to lightly push into a shrug. After a 10 second push of this sort the muscle will have released somewhat and slack can again be taken out as the head is repositioned, before a repetition of the procedure commences.

10. Assessment and treatment of scalenes

Assessment of shortness in scalenes (14)

The scalenes are prone to trigger point activity, and are a controversial muscle since they seem to be both postural and phasic, their status being modified by the type(s) of stress to which they are exposed (see Ch. 2, pp 28–29, for discussion of this topic). Janda reports that, 'spasm and/or trigger points are commonly present in the scalenes as also are weakness and/or inhibition.'

There is no easy test for shortness of the scalenes apart from observation, palpation and assessment of trigger point activity/tautness and a functional observation as follows:

- In most people who have marked scalene shortness there is a tendency to overuse these (and other upper fixators of the shoulder and neck) as accessory breathing muscles. There may also be a tendency to hyperventilation (and hence for there to possibly be a history of anxiety, phobic behaviour, panic attacks and/or fatigue symptoms). These muscles seem to be excessively tense in many people with chronic fatigue symptoms.
- The observation assessment consists of the operator placing their relaxed hands over the shoulders so that fingertips rest on the clavicles, at which time the seated patient is asked to inhale deeply. If the operator's hands noticeably rise towards the patient's ears during inhalation then there exists inappropriate use of scalenes, which indicates that they are stressed, which also means that, by definition, they will have become shortened and require stretching treatment.
- Alternatively, during the history taking inter-

view, the patient can be asked to place one hand on the abdomen just above the umbilicus and the other flat against the upper chest. On inhalation, the hands are observed and, if the upper one initiates the breathing process and rises significantly towards the chin, rather than moving forwards, a pattern of upper chest breathing can be assumed and therefore stress, and therefore shortness of the scalenes (and other accessory breathing muscles, notably sternomastoid).

Treatment of short scalenes by MET

Patient lies with head over the end of the table, supported on a cushion which rests on the knees of the seated operator (or the head can be supported by one cupped hand). The head/neck is in slight extension (painless) at this stage. The head is turned away from the side to be treated.

As with treatment of upper trapezius (p. 93) there are three positions of rotation required; a full rotation produces involvement of the more posterior fibres of the scalenes on the side from which the turn is being made (Fig. 4.20A); a half turn involves the middle fibres (Fig. 4.20B), and a position of only slight turn involves the more anterior fibres (Fig. 4.20C).

The operator's free hand is placed on the area just below the lateral end of the clavicle of the affected side (a side of hand, 'soft', contact is best, resting on the 2nd rib and upper sternal structures).

The patient is instructed, with appropriate breathing cooperation (see notes on breathing, Box 4.2, p. 67), to lift the forehead a fraction and to attempt to turn the head to the affected side, whilst resistance is applied preventing both

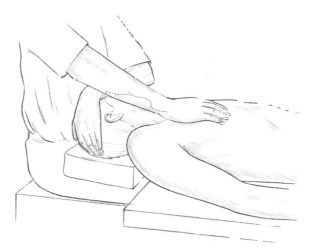

Figure 4.20B   MET treatment for the middle fibres of scalenes. The hand placement (thenar or hypothenar eminence of relaxed hand) is on the second rib below the centre of the clavicle.

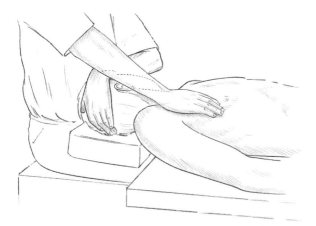

Figure 4.20A   MET for scalenus posticus. On stretching, following the isometric contraction, the neck is allowed to move into slight extension while a mild stretch is introduced by the contact hand which rests on the second rib, below the lateral aspect of the clavicle.

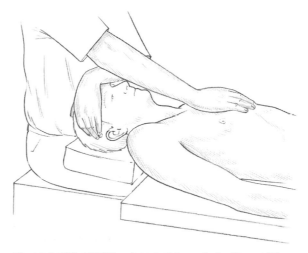

Figure 4.20C   MET treatment of the anterior fibres of the scalenes; hand placement is on the sternum.

movements ('lift and turn'). The effort, and therefore the counterpressure, should be modest and painless at all times.

After the 7 to 10 second contraction, the head is allowed to ease into extension and the contact hand on the 2nd rib and upper sternum pushes obliquely away towards the foot on that same side.

With the head half turned away from the affected side, the hand contact which applies the stretch into the middle fibres of the scalenes is just inferior to the middle aspect of the clavicle; and when the head is in the upright position, for anterior scalene stretch, the hand contact is on the upper sternum itself.

In all other ways the methodology is as described for the first position above.

CAUTION: It is important not to allow heroic degrees of neck extension during any phase of this treatment. There should be some neck extension, but it should be appropriate to the age and condition of the individual.

A degree of eye movement can assist scalene treatment. If the patient makes the eyes look downwards (towards the feet) and towards the affected side during the isometric contraction, she will increase the degree of contraction in the muscles. If, during the resting phase, when stretch is being introduced, she looks away from the treated side with eyes focused upwards towards the top of the head, this will enhance the stretch of the muscle.

This whole procedure should be performed several times, in each of the three positions of the head, for each side if necessary.

11. Assessment and treatment of sternocleidomastoid

Assessment for shortness of sternocleidomastoid (15)

As for the scalenes, there is no absolute test for shortness but observation of posture (hyper-lordotic neck, chin poked forward) and palpation of the degree of induration, fibrosis and trigger point activity can all alert to probable shortness of sternocleidomastoid (SCM). This is an accessory breathing muscle and, like the scalenes, will be shortened by inappropriate breathing patterns which have become habitual.

Observation is an accurate assessment tool. Since SCM is only just observable when normal, if the clavicular insertion is easily visible or any part of the muscle is prominent, this can be taken as a clear sign of tightness of the muscle.

If the patient's posture involves the head being held forward of the body, often accompanied by cervical lordosis and dorsal kyphosis (see Upper Crossed Syndrome notes in Chapter 2, p. 32), weakness of the deep neck flexors and tightness of SCM can be suspected.

An accurate functional test for shortness is observable by asking the supine patient to, very slowly, 'raise your head and touch your chin to your chest.' At the beginning of the movement of the head, as the patient lifts this from the table, the operator, standing to the side with head at the same level as the patient, would (if SCM were short) note that the chin was lifted first, allowing it to jut forwards, rather than the forehead leading the arc-like progression of the movement. In marked shortness of SCM the chin pokes forward in a jerk as the head is lifted. If the reading of this sign is unclear then Janda suggests that a slight resistance pressure be applied to the forehead as the patient makes the 'chin to chest' attempt. If SCM is short this will ensure the jutting of the chin at the outset, which is the sign of its shortness being sought.

Lewit suggests seeking indicative (of SCM shortness) painful points on the medial aspect of the clavicle and at the transverse process of the atlas, as well as trigger points in the muscle (sternal and clavicular divisions) and below the mastoid process.

Treatment of shortened SCM using MET

The patient is supine, head over the end of the table, supported by the operator's hand or on cushion on the lap of the seated operator, as in scalene treatment described above (p. 95).

Whereas in scalene treatment the instruction to the patient was to 'lift and turn' the rotated head/neck against resistance, in treating SCM the instruction is simply to lift the head. When the head is raised there is no need to apply resistance as gravity effectively does this.

After 7 to 10 seconds of this isometric contraction, and held breath, instruct the patient to release the effort (and breathe) and to allow the head/neck to return to a resting position in which some extension of the neck is allowed, while the soft edge of a hand applies oblique pressure/stretch to the sternum to take it away from the head towards the feet. The hand not involved in stretching the sternum away from the head should gently restrain the tendency the head will have to follow this stretch, but should under no circumstances apply pressure to stretch the head/neck while it is in this vulnerable position of slight extension. The degree of extension of the neck should be slight, 10° to 15° at most.

Maintain this stretch for some seconds to achieve release/stretch of hypertonic and fibrotic structures. Repeat as necessary.

Lewit advises a self-treatment manoeuvre in which the patient lies with the head just over the edge of a bed/table, rotated away from the treated side, chin resting on and supported by the edge of the bed, which acts as a fulcrum. The patient is taught to adopt this position, to take a deep held breath, to look towards the ceiling (with the eyes only – not by moving the head). This will slightly contract SCM on the side from which the patient has turned. On slow and full exhalation, the eyes look towards the floor and gravity will induce a stretch of the muscle as the top of the head eases towards the floor. Repeat this 2 or 3 times on each side of tightness in SCM.

Lewit reports that this technique provides very good results in treatment of restricted ('blocked') atlanto-occipital joint problems and relieves pain on the atlas transverse process.

12. Assessment and treatment of levator scapulae

Assessment of shortness in levator scapulae (16)

Test (a) A functional assessment involves applying the evidence we have seen (Ch. 2, p. 32) of the imbalances which commonly occur

between the upper and lower stabilisers of the scapula. In this process shortness is noted in pectoralis minor, levator scapulae and upper trapezius (as well as SCM) while weakness develops in serratus anterior, rhomboids, middle and lower trapezius – as well as the deep neck flexors.

Observation of the patient from behind will often show a 'hollow' area between the shoulder blades, where interscapular weakness has occurred, as well as an increased (over normal) distance between the medial borders of the scapulae and the thoracic spine, as the scapulae will have 'winged' away from it.

Test (b) To see this imbalance in action, Janda has the patient in the press-up position. On very slow lowering of the chest towards the floor from a maximum push-up position, the scapula(e) on the side(s) where stabilisation has been compromised will move outwards, laterally and upwards – into a winged position. This is diagnostic of weak lower stabilisers, which implicates tight upper stabilisers, including levator scapulae as inhibiting them.

Test (c) See assessment (b) for upper trapezius (p. 92) which needs to be only slightly modified as a palpation assessment for levator scapulae. The patient is supine with the neck flexed as well as being sidebent and rotated away from the side to be tested (for trapezius the head is rotated towards the side from which it was sidebent).

At this point the operator, standing at the head of the table, uses a contact on the shoulder (tested side) to assess the ease with which it can be depressed (moved distally). There should be an easy springing sensation as the shoulder is pushed towards the feet, with a soft end-feel to the movement. If there is a harsh, sudden end-point the levator scapula on that side is probably short.

Test (d) The patient lies supine with the arm of the side to be tested stretched out, with the hand and lower arm tucked under the buttocks, palm upwards, to help restrain movement of the shoulder/scapula.

The operator's arm is passed across and under the neck to rest on the shoulder of the side to be treated. The other hand supports the head.

With the forearm, the neck is lifted into full flexion (aided by the other hand) and is turned towards full sidebending and rotation away from the side to be tested.

Alternatively, the operator's knee can be used to sustain caudad pressure on the shoulder (locking the scapula) leaving two hands to achieve the desired head/neck position as described.

With the shoulder held caudad either by hand or knee, and the head/neck in the position described, there is a stretch on levator from both ends and if dysfunction exists and/or it is short, there will be marked discomfort reported at the insertion on the upper medial border of the scapula and/or pain near the spinous process of C2.

Test (e) Lewit achieves the same control by having the supine patient place their flexed elbow above their head, in contact with the operator's thigh or abdomen. This allows pressure through the long axis of the humerus to fix the scapula, while both hands are free to take the head/neck into its desired position.

Treatment using MET for levator scapulae

Method (a) The test position is used for treatment, either at the limit of easily reached range of motion, or well short of this, depending upon the degree of chronicity, which will also determine the degree of effort called for (20 to 30%) and the duration of each contraction (7 to 10 seconds, or up to 30 seconds) (see Fig. 4.21).

Following each isometric contraction, in which the patient has been asked to take their head back to the table, and slightly towards the side from which they have turned, against unmoving resistance (during a held breath), slack is taken out and the muscle is eased towards the new barrier (acute) or a little past the barrier, introducing passive stretch (if chronic). Repeat if needed. Caution is needed regarding stretching this somewhat sensitive area.

Method (b) If a stronger stretch is needed, the shoulder should be less firmly locked (remove hand from under buttock) and a slight shrug of the shoulder can accompany the head to table

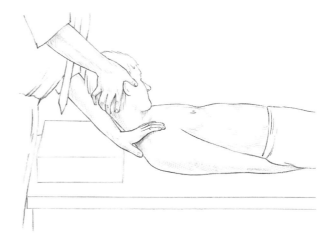

Figure 4.21 MET test (d) and treatment (a) position for levator scapula (right side).

instruction during the contraction, so involving both ends of the muscle. The stretch in this case would also involve some degree of pressure through the shoulder towards the feet.

13. Assessment and treatment of infraspinatus

Assessment of shortness in infraspinatus (17)

Test (a) The patient is seated, operator stands behind. The patient's arms are flexed at the elbow and held to the side, and the operator provides isometric resistance to external rotation of the lower arms (externally rotating them and also the humerus at the shoulder). If this effort is painful, an indication of probable infraspinatus shortening exists.

Test (b) The patient is asked to reach upwards, backwards and across to touch the upper border of their opposite scapula, so producing external rotation of the humeral head. If this effort is painful infraspinatus shortness is suspected.

Test (c) Visual evidence of shortness is obtained by having the patient supine, upper arm at right angles to the trunk, elbow flexed so that lower arm is parallel with the trunk, pointing caudad with the palm downwards. This brings the arm into internal rotation and places infraspinatus at stretch. If short, this will prevent the arm from being parallel with the

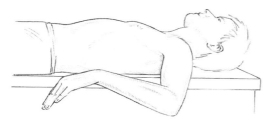

Figure 4.22   Assessment and self-treatment position for infraspinatus. If the upper arm cannot rest parallel to the floor, possible shortness of infraspinatus is indicated.

floor, obliging it to point somewhat towards the ceiling (see Fig. 4.22).

Treatment of infraspinatus using MET

The position described in test (c) above is adopted. The patient lifts the dorsum of the wrist towards the ceiling, against mild resistance from the operator for 10 seconds or so, and then, on relaxation, the wrist is taken towards the floor, so increasing internal rotation at the shoulder and stretching infraspinatus. Once again the rules as to acute and chronic con-

ditions should be kept in mind (see p. 66). Care needs to be taken to prevent the shoulder from rising from the table as rotation is introduced, so giving a false appearance of stretch in the muscle. A firm contact holding the shoulder to the table is advised (Fig. 4.23).

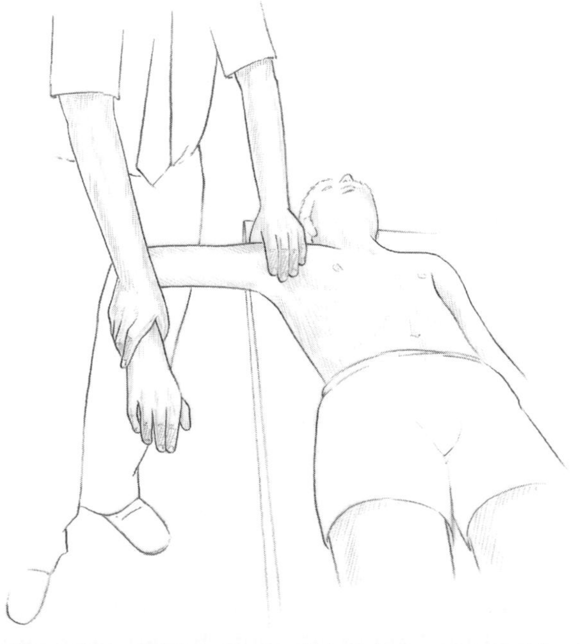

Self-treatment using gravity to stretch the muscle is advised by Lewit.

14. Assessment and treatment of subscapularis

Assessment for shortness in subscapularis (18)

Test (a) Direct palpation of subscapularis is required to define problems in it since pain patterns in the shoulder, arm, scapula and chest may all derive from it or from other sources.

The patient is supine. Operator grasps the affected side hand and applies traction while the fingers of the other hand palpate over the edge of latissimus dorsi in order to make contact with the ventral surface of the scapula, where subscapularis can be palpated. There may be a marked reaction from the patient when this is touched, indicating acute sensitivity.

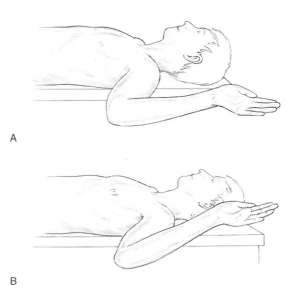

A

B

Figure 4.24A,B   Assessment and MET self-treatment position for subscapularis. If the upper arm cannot rest parallel to the floor, possible shortness of subscapularis is indicated.

Figure 4.23   MET treatment of infraspinatus. Note that the operator's left hand maintains a downward pressure to stabilise the shoulder to the table during this procedure.

Test (b) Subscapularis is an internal rotator of the arm and if it is restricted, reaching across to touch the opposite acromion may prove painful. This would also reduce its ability to externally rotate, as, for example, in the movement required to reach up and over behind the neck, to touch the opposite scapula.

Test (c) The patient is supine with the arm abducted to 90°, the elbow also at 90°, and the forearm in external rotation, palm upwards. The whole arm is resting at the restriction barrier, with gravity as its counterweight. If subscapularis is short the forearm will be unable to rest easily parallel with the floor but will be somewhat elevated (Fig. 4.24). Care is needed to prevent the anterior shoulder becoming elevated in this position (moving towards the ceiling) and so giving a false normal picture.[10]

Treatment of subscapularis using MET

The precise position is adopted as in assessment (c) above. The patient raises the forearm slightly against minimal resistance for 10 seconds and, following relaxation, gravity or slight assistance from the operator takes the arm into greater external rotation, to or through the barrier, as appropriate (acute/chronic). Repeat as necessary.

15. Assessment and treatment of supraspinatus

Assessment for shortness of supraspinatus (19)

Test (a) Supraspinatus shortness is assessed in a similar manner to infraspinatus. The patient is seated, operator stands behind. The patient's arm is flexed to 90° at the elbow and held to the side, and the operator provides isometric resistance to abduction of the lower arm by holding at the elbow. If this effort is painful, an indication of probable supraspinatus shortening exists.

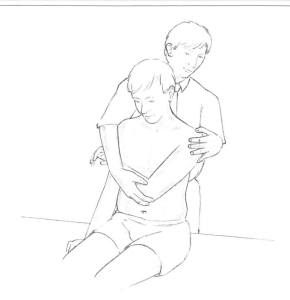

Figure 4.25 Position for test (b) and MET treatment (a) of supraspinatus.

Test (b) The operator stands behind the seated patient, with one hand stabilising the shoulder on the side to be assessed. The other hand reaches in front of the patient to support the flexed elbow/forearm (see Fig. 4.25). The patient's arm is adducted to the easy barrier and the patient attempts to abduct it. If pain (posterior shoulder area) is noted on this effort, this indicates supraspinatus dysfunction and probable shortness of the muscle.

Treatment of supraspinatus using MET

Method (a) The treatment position is as in test (b) above, except that the attempted abduction is resisted for 10 seconds after which the arm is adducted to or through the resistance barrier. Repeat as necessary.

Method (b) The patient is prone, with the affected side arm behind the back, elbow fully flexed, hand reaching across to the opposite side. The operator fixes the arm at or short of the resistance barrier, stabilising it at the elbow and the wrist, at which time a mild attempted abduction is made against firm resistance. After a 7 to 10 second contraction, the arm is taken into further adduction to or through the barrier. Repeat as necessary.

---

[10]There could be other reasons for a restricted degree of external rotation, and accurate assessment calls for direct palpation as in a) above.

16. Assessment and treatment of flexors of the arm

Assessment for shortness in flexors of the arm (20)

1. For shortness in biceps tendon

Test (a) Long biceps tendon is stressed if pain arises when the semi-flexed arm is raised against resistance.

Test (b) The patient fully flexes the elbow and the operator holds it in one hand while holding the patient's hand in the other. The patient is asked to resist as the operator attempts to externally rotate the elbow and to straighten the arm. If very unstable, the tendon may momentarily leave its groove and pain will result.

Test (c) The patient sits with extended arm (taking it backwards from the shoulder), half flexes the elbow so that dorsum of hand approximates the contralateral buttock. The patient attempts to flex the elbow further against resistance. If pain is noted, there is stress on the tendon and flexors are probably shortened.

MET treatment Lewit describes this method: The patient sits in front of the operator, with the affected arm behind the back, the dorsal aspect of that hand passing beyond the buttock on the opposite side. The therapist grasps this hand, bringing it into pronation, to take up the slack (see Fig. 4.26). The patient is instructed to attempt to take the hand back into supination. This is resisted for about 10 seconds by the operator, and the relaxation phase is used to take it further into pronation, with simultaneous extension of the elbow. Three to five repetitions may need to be performed. Self-treatment is possible, with the patient applying counter-pressure with the other hand.

2. Flexors of forearm: assessment for shortness and treatment using MET. Painful medial humeral epicondyle usually accompanies tension in the flexors of the forearm.

The patient is seated facing the operator, with flexed elbow supported by operator's fingers. Patient's hand is dorsiflexed at wrist, so that palm is upwards and fingers facing his own shoulder (see Fig. 4.27).

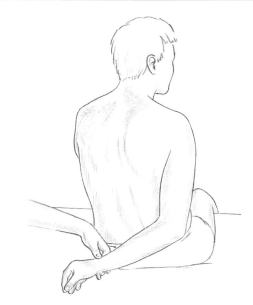

Figure 4.26   Assessment and MET treatment for dysfunction affecting biceps tendon.

Figure 4.27   Assessment and MET treatment for shortness of the flexors of the forearm.

The operator guides the wrist into greater flexion to an easy barrier, with pronation exaggerated by pressure on the ulnar side of the palm. This is achieved by means of the operator's thumb being placed on the dorsum of the patient's hand while the fingers stabilise the palmar aspect, fingertips pressing this towards the floor on the patient's ulnar side of the palm. The patient attempts to gently supinate the hand against resistance for 7 to 10 seconds following which, after relaxation, on an exhalation, dorsiflexion is increased to or through the new barrier (acute/chronic). Repeat as needed.

This method is easily capable of adaptation to self-treatment, by means of the patient applying the counterpressure.

3. Biceps brachii – assessment and MET treatment. If extension of the arm is limited, the flexors are probably short. Treatment of biceps brachii involves affected arm being held in extension at the easy barrier. The operator holds the patient's wrist in order to restrain a light effort to flex the elbow for 7 to 10 seconds after which, following appropriate rest and breathing cooperation (see notes on breathing, Box 4.2, p. 67), the arm is extended to or through the new resistance barrier (acute/chronic). Repeat several times.

17. Assessment and treatment of paravertebral muscles

Assessment of shortness in paravertebral muscles(21)

Test (a) The patient is seated on a treatment table, legs extended, pelvis vertical. Flexion is introduced in order to approximate forehead to knees. An even 'C' curve should be observed and a distance of about 4 in/10 cm from the knees achieved by the forehead. No knee flexion should occur and the movement should be a spinal one, not involving pelvic tilting (see Fig. 4.28).

Test (b) This assessment position is then modified to remove hamstring shortness from the picture, by having the patient sit at the end of the table, knees flexed over it. Once again the patient is asked to perform full flexion, without strain, so that forward bending is introduced to bring the forehead towards the knees. The pelvis should be fixed by the placement of the patient's hands on the pelvic crest.

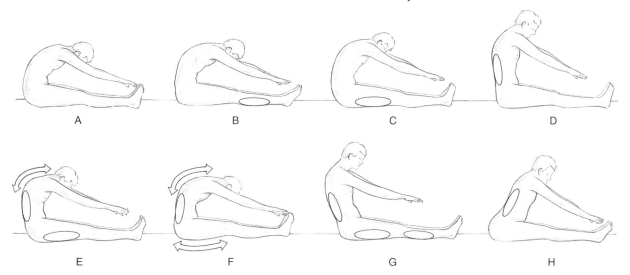

Figure 4.28 Tests for shortness of the erector spinae and associated postural muscles.
A Normal length of erector spinae muscles and posterior thigh muscles.
B Tight gastrocnemius and soleus; the inability to dorsiflex the feet indicates tightness of the plantar-flexor group.
C Tight hamstring muscles, which cause the pelvis to tilt posteriorly.
D Tight low back erector spinae muscles.
E Tight hamstrings; slightly tight low back muscles and overstretched upper back muscles.
F Slightly shortened lower back muscles, stretched upper back muscles and slightly stretched hamstrings.
G Tight low back muscles, hamstrings and gastrocnemius/soleus.
H Very tight low back muscles, with lordosis maintained even in flexion.

If bending of the trunk is greater in this position than in (a) above, then there is probably shortened hamstring involvement. During these assessments, areas of shortening in the spinal muscles may be observed as 'flat', or even, in the lumbar area, of a reversed curve.

For example, on forward bending a lordosis may be maintained in the lumbar spine, or flexion may be very limited even without such lordosis. There may be evidence of obvious overstretching of the upper back and relative tightness of the lower back.

All areas of 'flatness' are charted since these represent an inability of those segments to flex, which involves the erector spinae muscles as a primary or a secondary feature. If the flexion restriction relates to articular factors, the erector group will nevertheless benefit from MET. If they are primary causes of the flexion restriction then MET attention is even more indicated.

Lewit points out that patients with a long trunk and short thighs may perform the movement without difficulty, even if the erectors are short, whereas if the trunk is short and the thighs long, even if the erectors are supple, flexion will not allow the head to approximate the knees.

In the modified position, with patient's hands on the crest of the pelvis, and the patient 'humping' her spine, he suggests observation of the presence or otherwise of lumbar kyphosis for evidence of shortness in that region. If it fails to appear, erector spinae shortness in the lumbar region is likely. This, together with the presence of flat areas, provides significant evidence of shortness.

Test (c) Once all flat areas are noted and charted, the patient is placed in prone position. The operator squats at the side and observes the spinal 'wave' as deep breathing is performed. There should be a wave of movement starting at the sacrum and finishing at the base of the neck on inhalation. Areas of restriction ('flat areas'), lack of movement, or where motion is not in sequence, should be noted and compared with findings from tests (a) and (b) above.

Periodic review of the relative normality of this wave is a useful guide to progress (or lack of it) in normalisation of the functional status of the respiratory and spinal structures.

MET treatment of erector spinae muscle

The patient sits with back to operator on treatment couch, legs hanging over side and hands clasped behind the neck. The operator places knee on the couch close to the patient, at the side towards which sidebending and rotation will be introduced. The operator passes a hand in front of the patient's axilla on the side to which the patient is to be rotated, across the front of the patient's neck, to rest on the shoulder opposite. The patient is drawn into flexion, sidebending and rotation over the operator's knee. The operator's free hand monitors the area of tightness (as evidenced by 'flatness' in the flexion test) and ensures that the various forces localise at the point of maximum contraction/tension. When the patient has been taken to the comfortable limit of flexion, she is asked to either look (eyes only) towards the direction from which rotation has been made, whilst holding the breath for 7 to 10 seconds, or to do this while also introducing a very slight degree of effort towards rotating back to the upright position, against firm resistance from the operator. (See Fig. 6.1B, p. 120; also Figs 6.1E,F and 6.2A,B, pp. 121–122.)

It is useful to have the patient 'breathe into' the tight spinal area which is being palpated and monitored by the operator. This will cause an additional increase in isometric contraction of the muscles which have shortened.

The patient is then asked to release the breath, completely relax and to look towards the direction in which sidebending/rotation is being introduced (i.e. towards the resistance barrier). The operator waits for the patient's second full exhalation and then takes the patient further in all the directions of restriction, towards the new barrier, not through it.

This whole process is repeated several times, at each level of restriction/flatness.

At the end of each sequence of repetitions the patient may be asked to breathe in and to gently attempt to rotate further against resistance, towards the restriction barrier, whilst holding the breath for 7 to 10 seconds. This involves contraction of the antagonists. After relaxation, the new barrier is again approached.

Thoracolumbar dysfunction

This important transition region was discussed briefly in the section dealing with quadratus lumborum (p. 86), and deserves special attention due to its particularly vulnerable 'transition' status involving the powerful effect that spasm and tightness of the major stabilising muscles of the region can have on it; notably, psoas, the thoracolumbar erector spinae and quadratus lumborum, as well as the influence of rectus abdominus in which weakness is all too common (see Lower Crossed Syndrome notes in Ch. 2, p. 33).

Screening for lumbodorsal dysfunction involves having the patient straddle the couch (so locking the pelvis) in a slightly flexed posture (slight kyphosis). Rotation in either direction enables segmental impairment to be observed at the same time as the spinous processes are monitored. Restriction of rotation is the most common characteristic of this dysfunction.

MET treatment of thoracolumbar dysfunction

Psoas and/or quadratus lumborum may be treated as above.[11]

Assessment for shortness in erector spinae muscles of the neck (21c)

The patient is supine and the operator stands at the head of the table, or to the side, supporting the neck structures in one hand and the base of the skull in the other, to afford complete support for both.

When the head/neck is lifted into flexion the chin should easily be able to be brought into contact with the suprasternal area, without force.

If there remains a noticeable gap between the tip of the chin (ignore double chin tissues) and the upper chest wall, then the neck extensors are considered to be short.

Treatment of short neck extensor muscles using MET

The neck of the supine patient is flexed to its easy barrier of resistance or short of this (acute/chronic) and the patient is asked to extend the neck (take it back to the table) using minimal effort on an inhalation, against resistance. If the hand positions as described in the test above are not comfortable, then try placing the hands, arms crossed, so that a hand rests on each shoulder, or upper anterior shoulder area, while the head rests on the crossed forearms.

After the contraction, the neck is flexed further to, or through, the barrier of resistance, as appropriate (acute/chronic).

A further aid during the contraction phase is to have the operator contact the top of the head with his abdomen and to use this contact to prevent the patient tilting the head upwards. This allows for an additional isometric contraction which involves the short extensor muscles at the base of the skull. The subsequent stretch, as above, will involve these as well. Repetitions of the stretch should be performed until no further gain is possible, or until the chin easily touches the chest on flexion.

No force should be used, or pain produced during this procedure.[12]

---

[11]Not all the muscles involved in thoracolumbar dysfunction pattern described above may need treatment since when one or other is treated appropriately the others tend to normalise. Underlying causes must also always receive attention.

[12]If assessment is being made of cervical rotational efficiency/restriction a simple screening device is available. When the head/neck is in full flexion, and rotation is introduced, all rotation below C3 is blocked. Therefore, if rotation is tested in full flexion and there is a limitation to one side, this probably represents a problem in the atlanto-occipital or atlanto-axial joints. When the head is fully extended on the neck, then the atlanto-occipital and C2/3 joints are locked and any rotation restriction relates to problems below that level.

A variety of MET treatment methods for pelvic and spinal joint restrictions, including the cervical area, will be found in Chapter 6, and these can be used alongside the more general, muscle orientated approaches detailed above.

REFERENCES

Basmajian J 1974 Muscles alive. Williams and Wilkins, Baltimore

Cailliet R 1962 Low back pain syndrome. Blackwell, Oxford

Chaitow L 1991 Soft tissue manipulation. Healing Arts Press, Rochester

Dvorak J, Dvorak V 1984 Manual medicine – diagnostics. George Thiem Verlag, New York

Evjenth O 1984 Muscle stretching in manual therapy. Alfta Rehab, Alfta, Sweden

Fryette 1954 Principles of osteopathic technic. Yearbook of the Academy of Applied Osteopathy 1954

Greenman P 1989 Principles of manual medicine. Williams and Wilkins, Baltimore

Janda V 1983 Muscle function testing. Butterworths, London

Janda V 1988 In: Grant R (ed) Physical therapy of the cervical and thoracic spine. Churchill Livingstone, New York

Lewit K 1985a Muscular and articular factors in movement restriction. Manual Medicine 1: 83–85

Lewit K 1985b Manipulative therapy in rehabilitation of the motor system. Butterworths, London

Lewit K 1992 Manipulative therapy in rehabilitation of the locomotor system. Butterworths, London

Mennell J 1964 Back pain. T and A Churchill, Boston

Moore M et al 1980 Electromyographic investigation manual of muscle stretching techniques. Medicine and Science in Sports and Exercise 12: 322–329

Retzlaff E et al 1974 The piriformis muscle syndrome. Journal of the American Osteopathic Association 173: 799–807

Richard R 1978 Lesions osteopathiques du sacrum. Maloine, Paris

Rolf I 1977 Rolfing – integration of human structures. Harper and Row, New York

Ruddy T 1962 Osteopathic rapid rhythmic resistive technic. Academy of Applied Osteopathy Yearbook 1962: 23–31

Stiles E 1984 Manipulation – a tool for your practice? Patient Care 18: 16–42

TePoorten B 1960 The piriformis muscle. Journal of the American Osteopathic Association 69: 150–160

Williams P 1965 Lumbosacral spine. McGraw Hill, New York

5

Manual resistance techniques and rehabilitation

Craig Liebenson DC

Manual Resistance Techniques (MRTs) were introduced in the late 1940s by physical therapists as part of the proprioceptive neuromuscular facilitation philosophy, and by osteopaths as muscle energy procedures. MRTs are powerful rehabilitation tools which can be used to relax or inhibit a muscle, stretch muscle or fascia, facilitate a muscle, or mobilise a joint. They are ideal bridges between passive and active care. MRTs are the workhorses of rehabilitation because of their effectiveness and versatility.

FUNCTIONAL PATHOLOGY OF THE MOTOR SYSTEM

Muscular imbalance

There are a number of possible treatment targets which can be addressed with MRTs. Muscle spasm, increased muscle tension, muscle stiffness, connective tissue adhesions, muscle inhibition, muscle atrophy, and joint dysfunction are all amenable to treatment by MRTs.

Dysfunction in the locomotor system is rarely an isolated situation. Typically, any dysfunction or pathology is part of a general 'chain reaction' involving some 'key link' and numerous compensations or adaptations. Muscular imbalance between antagonist muscles is a classic expression of this chain of events. The muscle imbalance usually involves weakness or at least inhibition of an important agonist (i.e. gluteus maximus for hip extension) along with tightness or tension in antagonist (i.e. psoas) or synergist (i.e. hamstring, erector spinae) muscles. This muscular imbalance causes an altered movement pattern (i.e. hip extension) which is clinically relevant for its relationship to certain fundamental daily skills (i.e. toe off during gait-propulsion). Muscle imbalances cause altered movement patterns and will invariably contribute to painful dysfunctions (i.e. trigger points, joint instability) or structural pathology (disc/nerve root syndrome, arthritis).

The muscle imbalance itself is at the crossroads of the body's adaptation to any asymptomatic muscular or articular dysfunction or to an injury. Addressing muscle imbalance and improving altered movement patterns is an essential component to improving biomechanical and kinesiological integrity of the locomotor system.

Terminology

Muscles are often said to be short, tight, tense, or in spasm. These terms, however, are used very loosely. In order to provide proper indications for the use of MRTs we should define our treatment objectives. Muscles suffer either neuromuscular, viscoelastic, or connective tissue alterations. A tight muscle could have either increased neuromuscular tension or connective tissue fibrosis.

Spasm

Muscle spasm is a neuromuscular phenomenon relating either to an upper motor neuron disease, or an acute reaction to pain or tissue injury. Electromyographic (EMG) activity is increased in these cases. Examples include spinal cord injury, reflex spasm such as appendicitis or acute lumbar antalgia with loss of flexion relaxation response (Triano & Schultz 1987). Long-lasting noxious stimulation has been shown to activate the flexion withdrawal reflex (Dahl et al 1992).

Tension without EMG elevation

Increased muscle tension can occur without a consistently elevated EMG. An example is trigger points, in which case a muscle fails to relax properly. Muscles housing trigger points have been shown to have dramatically different levels of EMG activity within the same functional muscle unit. Hubbard and Berkoff (1993) showed EMG hyperexcitability in the nidus of the trigger point in a taut band which had a characteristic pattern of reproducible referred pain.

Increased stretch sensitivity

Increased sensitivity to stretch can also lead to increased muscle tension. This has been shown to occur under conditions of local ischaemia (Mense 1993). Low intensity stimulation of joint afferents in the knee has been shown to influence stretch sensitivity (Johansson et al 1988, 1989). Schiable and Grubb (1993) have implicated reflex discharges from joints mediated by efferent sympathetic fibres in the production of such neuromuscular tension. Johannson et al (1991) have correlated non-noxious joint afferent activity with increased reflex gamma-motorneuron activity. According to Janda (1991) neuromuscular tension can also be increased by central influences due to limbic dysfunction.

Viscoelastic influence

Muscle stiffness is a viscoelastic phenomena described by Walsh (1992). This has to do with fluid mechanics and viscosity of tissue and is not a neuromuscular phenomenon. Fibrosis occurs in muscle or fascia gradually and is typically related to post-trauma adhesion formation. Lehto (1986) found that fibroblasts proliferate in injured tissue during the inflammatory phase. If the inflammatory phase is prolonged, then a connective tissue scar will form as the fibrosis is not absorbed.

Trigger point influence

Various studies have demonstrated that trigger points in one muscle are related to inhibition of another functionally related muscle (Simons 1993, Headley 1993). In particular, it was shown by Simons that the deltoid muscle can be inhibited when infraspinatus trigger points are present (Simons 1993). Headley has shown that lower trapezius inhibition is related to trigger points in the upper trapezius (Headley 1993).

Atrophy

In chronic back pain patients, generalised atrophy has been found, but with a relative increase in the cross-sectional analysis of muscles on the symptomatic side (Stokes et al 1992). Type I (postural or aerobic) fibre hypertrophy on the symptomatic side and type II (phasic or anaerobic) fibre atrophy bilaterally has been documented in chronic back pain patients (Fitzmaurice et al 1992).

What is weakness?

Muscle weakness is another term that is used loosely. A muscle may only be inhibited, meaning that it has not suffered disuse atrophy, but is weak due to a reflex phenomenon. Inhibited muscles are capable of spontaneous strengthening when the inhibitory reflex is identified and remedied (usually through soft tissue or joint manipulation by MRT or other means).

A typical example is reflex inhibition from an antagonist muscle due to Sherrington's law of reciprocal inhibition. Reflex inhibition of the vastus medialis oblique (VMO) muscle after knee inflammation/injury has been repeatedly demonstrated (DeAndrade et al 1965, Brucini et al 1981, Spencer et al 1984). Hides et al (1994) have found unilateral, segmental wasting of the multifidus in acute back pain patients. This occurred rapidly and thus was not considered to be a disuse atrophy.

True muscle weakness is a result of lower motor neuron disease (i.e. nerve root compression), or disuse atrophy. In chronic back pain patients, generalised atrophy has been demonstrated (Stokes et al 1992). This atrophy is selective in the type II (phasic) muscle fibres bilaterally (Fitzmaurice et al 1992).

Various pathologic situations have been mentioned above which can effect either the flexibility or the strength of muscles. The result is muscular imbalance involving increased tension or tightness in postural muscles, along with

Table 5.1   Scientific evidence of muscular imbalance

Documented evidence of increased neuromuscular tension or connective tissue changes:

1. Relative type I (postural) muscle fibre hypertrophy on symptomatic side in chronic low back pain (Stokes et al 1992, Fitzmaurice et al 1992)
2. Prolonged nociceptive bombardment can lead to the flexion reflex due to excessive contraction of skeletal muscles in the vicinity of the nociceptors (Bromm 1994)
3. Lehto found that fibroblasts proliferate in injured tissue during the inflammatory phase (Lehto et al 1986)

Documented evidence of muscle inhibition, weakness or atrophy:

1. Reflex inhibition of VMO after knee inflammation/injury (DeAndrade et al 1965, Brucini et al 1981, Spencer et al 1984)
2. Unilateral, segmental type II (phasic) muscle fibre atrophy after onset of acute low back pain (Hides et al 1994)
3. Bilateral, type II (phasic) muscle fibre atrophy in chronic low back pain (Stokes et al 1992, Fitzmaurice et al 1992)

Adapted with permission from C L Liebenson Rehabilitation of the spine: a practitioner's manual. Williams and Wilkins, Baltimore (in press) 1995.

inhibition or weakness of phasic muscles. This evidence is summarised in Table 5.1.

ALTERED MOVEMENT PATTERNS AND JOINT OVERSTRESS

It is important to understand that nearly any tissue can generate pain. That being said, we should strive to uncover the primary dysfunctions and not the compensations or adaptations which inevitably surround the area of pain. Biomechanically, it is evident that there are functional chains in the body. The lower extremity in human beings functions as a closed kinetic chain. Any dysfunction in the foot, such as hyperpronation, will inevitably lead to a chain reaction involving the knee, hip and lumbar spine. A stiff hip joint may develop as a result of hip flexor tightness and this may lead to compensatory lumbar hypermobility and paraspinal trigger points. The result may be low back or buttock pain or, even worse, a lumbosacral nerve root syndrome. In the case of back or buttock pain, local treatments involving manipulative (joint or soft tissue) therapy may improve the situation.

However, to prevent recurrences or to treat the pain if it is chronic, usually requires a more comprehensive approach. Treatment aimed at relaxing a tight psoas and strengthening a weak gluteus maximus may be the primary treatment for lumbosacral facet pain or paraspinal myofascial pain.

Altered movement patterns

When pain, muscular imbalance, trigger points, or joint dysfunction are present, the patient's performance of certain stereotypical movement patterns will be altered. The negative relationship between various individual functional pathologies and the abnormal performance of basic movement patterns is self-perpetuating. Altered or faulty movement patterns themselves place new strains on the locomotor system and lead to the spread of a local problem beyond a single region.

Trick patterns

Such movement patterns were first recognised clinically by Janda, when it was noticed that classic muscle testing methods did not differentiate between normal recruitment of muscles for an action and 'trick' patterns of substitution during an action (Janda 1978). Traditional muscle testing was only interested in resistance and strength, not in coordination. However, so-called 'trick' movements are uneconomical and place unusual strain on the joints. They involve muscles in uncoordinated ways and are related to poor endurance. On a traditional test of prone hip extension, it is difficult to identify overactivity of the lumbar erector spinae or hamstrings as substitutes for an inhibited gluteus maximus. Tests developed by Janda are far more sensitive and allow us to identify muscle imbalances, faulty movement patterns and joint overstrain by seeing abnormal substitution during our muscle testing protocols (Janda 1978).

Joint implications

When a movement pattern is altered, the activation sequence or firing order of different muscles involved in a specific movement is disturbed. The prime mover may be slow to activate, while synergists or stabilisers substitute and become overactive. When this is the case, new joint stresses will be encountered. Sometimes the timing sequence is normal but the overall range may be limited due to joint stiffness or antagonist muscle shortening.

Observation or palpation?

Key altered movement patterns that can be tested as part of a screening examination for locomotor dysfunction are shown in Figures 5.1 to 5.6.

In general, observation (from a few feet away) alone is all that is needed to determine the altered movement pattern. However, light palpation may also be used if observation is difficult due to poor lighting, a vision problem, or if the patient is not sufficiently disrobed.

Headache patients as examples

Support for this hypothesis has emerged in the evaluation of headache patients. Key findings which differentiate headache from non-headache sufferers have been found to include:

- A forward head posture
- Decreased isometric strength of neck flexors
- Decreased isometric endurance of neck flexors (Watson & Trott 1993, Treleaven et al 1994).

Neck flexion in these patients is accomplished by overactivity of the superficial muscles (sternocleidomastoid and scalene), while the deep muscles (longus colli and capitus) are underactive. Furthermore, shortening of the suboccipitals is also present.

Impaired hip extension movement pattern

Another classic example of muscular imbalance causing an impaired movement pattern occurs when tight hip flexors are combined with a weak gluteus maximus, thus causing an inefficient or uncoordinated hip extension movement pattern.

The gluteus maximus may be inhibited and activate poorly during the movement, leading to overactivity of the stabilisers in the lumbar spine, the erector spinae muscles. While such an altered pattern may have formed as a result of hip flexor compensation to a low back strain, hyperpronation problem, leg length inequality, etc. it will eventually perpetuate instability by overstressing the lumbar joints on its own. In this case the following functional pathology will all be interconnected:

- Shortening of the psoas
- Inhibition/weakness/trigger points of the gluteus maximus
- Overactivity/trigger points of the lumbar erector spinae
- Lumbar spine joint dysfunction
- Altered coordination/endurance of hip extension, particularly during gait.

Integrated movement

Testing individual muscles for strength without concern for the speed of activation, or the activation sequence of agonist, synergists, and stabilisers is an error.

According to Korr (1976), 'The brain thinks in terms of whole motions, not individual muscles.' Muscles may have anatomical individuality, but they function interdependently to create smooth, well-orchestrated movements.

Compensation

Altered movement patterns often develop as a compensation to pain or injury. Herring (1990) said, 'signs and symptoms of injury abate, but these functional deficits persist … adaptive patterns develop secondary to the remaining deficits … rehabilitation … must therefore address more than the pain of an individual injury. The rehabilitation plans … must be oriented toward return to function, not just relief of symptoms.'

CLINICAL APPLICATION

Manual resistance techniques (MRTs) are an essential tool in addressing muscular imbalance

Table 5.2   MRT goals and techniques

1. Muscle inhibition/relaxation – gentle isometric or concentric MRT (may be used with or without antagonist contraction)
2. Muscle or fascial stretch – gentle eccentric or hard concentric MRT
3. Muscle facilitation – isometric, concentric, and eccentric MRT (numerous repetitions are optimal)

Adapted with permission from C L Liebenson Rehabilitation of the spine: a practitioner's manual. Williams and Wilkins, Baltimore (in press) 1995.

and altered movement patterns. Table 5.2 reviews specific goals which can be accomplished via MRTs.

The facet syndrome

The facet joint is a common culprit as the source of an individual's low back pain. Pain is usually worse with weight bearing, walking, and back bending. From a rehabilitation perspective, facet syndromes are likely to be correlated with the altered movement patterns of hip extension and trunk flexion.

An altered hip extension synergy (identified using the prone test – see Fig. 5.1) will lead to overstress of the lumbar facets. This is a result of compensatory hyperextension (and hypermobility) of the lumbar spine for a hypomobile hip joint or weak gluteus maximus muscle (Jull & Janda 1987, Lewit 1991).

An altered trunk flexion movement pattern (abnormal substitution of psoas for weak abdominals – see Fig. 5.2) will overstress the L5/S1 joints due to poor lumbopelvic stabilisation (Janda 1978, Jull & Janda 1987, Lewit 1991).

Trunk stabilisation through improved co-contraction of abdominals and gluteals will be essential to prevent recurrences.

Treatment protocol

- Joint mobilisation/manipulation lumbar spine and hip joint
- Relaxation/stretch of hip flexors and erector spinae
- Facilitate/strengthen gluteus maximus and rectus abdominus.

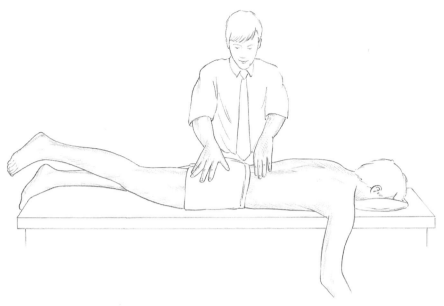

Figure 5.1 Hip extension test. The normal activation sequence is gluteus maximus and hamstrings, followed by erector spinae (conlateral then ipsilateral).

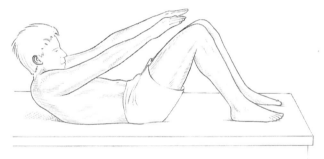

Figure 5.2 Trunk flexion test.
Figure 5.2A Normal – ability to raise trunk until scapulae are off table without feet lifting off or low back arching.

Figure 5.2B Abnormal – when feet rise up or low back arches before scapulae are raised from the table.

The sacroiliac syndrome

Sacroiliac (SI) joints, when irritated, can cause local or referred pain to the buttock or posterior thigh. SI pain is worse with weight bearing (standing or sitting) and walking. Passive tests such as Gaenslen's manoeuvre are often confirmatory as a result of pain provocation.

Gaenslen's test

This involves flexing one hip while simultaneously extending the other off the side of the table (patient lies supine close to edge of table which allows extension of hip on that side). The resultant anterior rotation of the ilium on the sacrum on the side of the extended hip causes pain in a sensitive SI joint on that side.

SI syndromes are intertwined with altered hip abduction. Weakness/inhibition of the gluteus medius will result in increased lateral shearing forces across the pelvis. Typically during hip abduction testing, if the gluteus medius is weak, hip flexion during abduction will be seen as a result of tensor fascia lata (TFL) overactivity (see Fig. 5.3) (Jull & Janda 1987, Lewit 1991).

Often, overactivity or shortening of the antagonistic hip adductors will add torsion to the pelvis through the pubic bones, and thus aggravate an SI joint problem. The gluteus medius insufficiency is also often combined with piriformis overactivity which can neutralise the gluteus medius and lead to lateral instability. The quadratus lumborum (QL) may substitute for gluteus medius weakness with resultant excessive hip hiking during gait and myofascial trigger point formation. A rehabilitation solution is the only way to crack this code.

Treatment protocol

- Mobilise/manipulate SI joint
- Relaxation/stretch – adductors, piriformis, QL, TFL
- Facilitate/strengthen – gluteus medius.

Headache

Headaches can derive from a variety of sources – cervical, myofascial, nutritional, vascular (migraine), and other (cluster). Cervical and myofascial headaches are quite amenable to rehabilitation.

Two postural syndromes are common. One is the imbalance between overactive/tight upper trapezius and levator scapulae with inhibited/weak lower and middle trapezius (Lewit 1991, Janda 1986). These headache patients have most

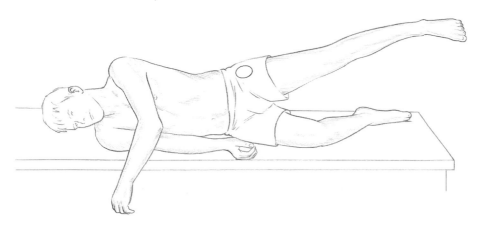

Figure 5.3A   Normal – hip abduction to 45°.

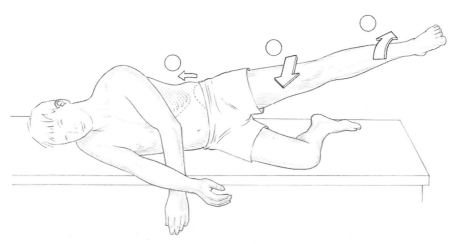

Figure 5.3B   Abnormal – if hip flexion, external rotation or 'hiking' occurs, or pelvic rotation takes place during hip abduction.

Figure 5.3   Hip abduction test.

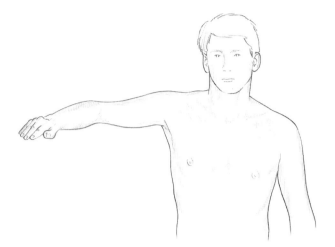

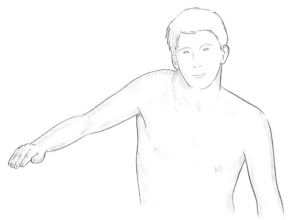

Figure 5.4   Scapulo-humeral rhythm test.
Figure 5.4A   Normal – elevation of shoulder after 60° of arm abduction.

Figure 5.4B   Abnormal – if elevation of the shoulder or winging of the scapulae occurs within the first 60° of shoulder abduction.

of their pain in the back of the head, but occasionally over the eyes. Loss of the normal scapulohumeral rhythm is confirmation of this muscle imbalance (see Fig. 5.4).

The more severe headache patients (usually women) and even some migraine sufferers who present with forehead and eye pain, will typically have weakness of their deep neck flexors and loss of lower cervical extension (Watson & Trott 1993, Treleaven et al 1994, Lewit 1991, Janda 1986). These patients present with an identifiable head forward posture. On neck flexion from the supine position, the chin cannot be held in a tucked position as overactivity in the sterno-

cleidomastoid (SCM) causes cervicocranial hyperextension (see Fig. 5.5). Chest breathing with overactivity of the scalenes is also commonly seen in headache sufferers (Lewit 1991).

Treatment protocol

- Mobilise/manipulate cervicocranial and cervicothoracic junctions
- Relaxation/stretch – upper trapezius, levator scapulae, suboccipitals, SCM
- Facilitate/strengthen – middle and lower trapezius, serratus anterior
- Retrain diaphragmatic respiration.

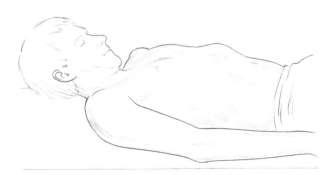

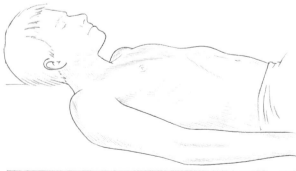

Figure 5.5   Neck flexion test.
Figure 5.5A   Normal – ability to hold chin tucked while flexing the head/neck.

Figure 5.5B   Abnormal – if the chin pokes forwards while attempting head flexion.

Thoracic outlet syndrome

Shoulder blade pain, chest pain, cervicobrachial syndromes, and headaches can all come from a forward drawn head and downward shift of the cervicothoracic junction (Lewit 1991, Janda 1986, Liebenson 1988, Matthews 1986). These symptoms are common, especially in women (in part due to the forward pull of the breasts) and men who have either poor posture or who, through weight lifting, have overdeveloped their anterior chest and neck musculature relative to their upper back muscles. It has been demonstrated that scalene and pectoralis minor muscle entrapment of the cervicobrachial plexus can precipitate arm pain and dysthesia along an ulnar distribution (Ribbe & Lindgren 1986, Roos 1979).

Restoring muscle balance to the anterior and posterior chest/upper back muscles will be the key to successfully managing this common problem. A forward drawn posture, rounded shoulders, head forward, and winging of the scapulae will be key examination findings. Weakness of the lower fixators of the scapulae (serratus anterior, lower and middle trapezius) and overactivity of the pectorals and upper fixators of the scapulae (upper trapezius and levator scapulae) will commonly be found. Trunk lowering from a push-up position can be used as a screening test for serratus anterior weakness (see Fig. 5.6).

A superior shift, or 'winging', of the scapulae will be positive test results. Additionally, tightness of the superficial anterior neck muscles (scalene and SCM) and suboccipitals will be found along with weakness of the deep neck flexors.

Treatment protocol

- Mobilise/manipulate upper thoracic and lower cervical spine
- Relaxation/stretch – pectoralis major and minor, scalenes, suboccipitals
- Facilitate/strengthen – middle and lower traps, serratus anterior.

Figure 5.6 Abnormal trunk lowering from a push-up test. The scapulae should protract without winging, or shifting superiorly, during trunk lowering.

REFERENCES

Bromm B 1994 Corticalization of chronic pain. APS 3: 131–135

Brucini M, Duranti R, Galleti R, Pantaleo T, Zucchi P L 1981 Pain thresholds and electromyographic features of periarticular muscles in patients with osteoarthritis of the knee. Pain 10: 57–66

Dahl J B, Erichsen C J, Fuglsang-Frederiksen A, Kehlet H 1992 Pain sensation and nociceptive reflex excitability in surgical patients and human volunteers. British Journal of Anaesthesia 69: 117–121

DeAndrade J R, Grant C, Dixon A St J 1965 Joint distension and reflex muscle inhibition in the knee. Journal of Bone and Joint Surgery 47: 313–322

Fitzmaurice R, Cooper R G, Freemont A J 1992 A histo-morphometric comparison of muscle biopsies from normal subjects and patients with ankylosing spondylitis and severe mechanical low back pain. Journal of Pathology 163: 182

Headley B J 1993 Muscle inhibition. Physical Therapy Forum 24:November 1

Herring S A 1990 Rehabilitation of muscle injuries. Medicine and Science in Sports and Exercise 22: 453, 456

Hides J A, Stokes M J, Saide M, et al 1994 Evidence of lumbar multifidus muscles wasting ipsilateral to symptoms in patients with acute/subacute low back pain. Spine 19: 165–172

Hubbard D R, Berkoff G M 1993 Myofascial trigger points show spontaneous needle EMG activity. Spine 18: 1803–1807

Janda V 1978 Muscles, central nervous motor regulation, and back problems. In: Korr I M (ed) Neurobiologic mechanisms in manipulative therapy. Plenum, New York

Janda V 1986 Some aspects of extracranial causes of facial pain. Journal of Prosthetic Dentistry 56: 484

Janda V 1991 Muscle spasm – a proposed procedure for differential diagnosis. Manual Medicine 6: 136–139

Johansson H, Sjolander P, Sojka P 1988 Fusimotor reflexes in triceps surae muscle elicited by natural and electrical stimulation of joint afferents. Neuro-Orthop. 6: 67–80

Johansson H, Sjolander P, Sojka P, Wadell I 1989 Reflex actions on the gamma-muscle spindle systems acting at the knee joint elicited by the stretch of the posterior cruciate ligament. Neuro-Orthop 8: 9–21

Johansson H, Sjolander P, Sojka P 1991 Receptors in the knee joint ligaments and their role in the biomechanics of the joint. Critical Reviews in Biomedical Engineering 18: 341–368

Jull G, Janda V 1987 Muscles and motor control in low back pain. In: Twomney L T, Taylor J R (eds) Physical therapy for the low back. Clinics in Physical Therapy, Churchill Livingstone, New York

Korr I 1976 The spinal cord as organizer of disease processes: some preliminary perspectives. JAOA 76: 35–45

Lehto M, Jarvinen M, Nelimarkka O 1986 Scar formation after skeletal muscle injury. Archives of Orthopaedic and Trauma Surgery 104: 366–370

Lewit K 1991 Manipulative therapy in rehabilitation of the motor system, 2nd edn. Butterworths, London

Liebenson C S 1988 Thoracic outlet syndrome. JMPT 11: 6

Matthews M 1986 The T4 syndrome. Australian Journal of Physiotherapy 32(2): 123–124

Mense S 1993 Nociception from skeletal muscle in relation to clinical muscle pain. Pain 54: 241–290

Ribbe E B, Lindgren S H S 1986 Clinical diagnosis of TOS. Manual Medicine 2: 82–85

Roos D B 1979 New concepts of TOS that explain etiology, symptoms, diagnosis, and treatment. Vascular Surgery 13: 313–321

Schiable H G, Grubb B D 1993 Afferent and spinal mechanisms of joint pain. Pain 55: 5–54

Simons D G 1993 Referred phenomena of myofascial trigger points. In: Vecchiet L, Albe Fessard D, Lindlom U New trends in referred pain and hyperalgesia. Elsevier, Amsterdam

Spencer J D, Hayes K C, Alexander I J 1984 Knee joint effusion and quadriceps reflex inhibition in man. Archives of Physical Medicine and Rehabilitation 65: 171–177

Stokes M J, Cooper R G, Jayson M I V 1992 Selective changes in multifidus dimensions in patients with chronic low back pain. Eur Spine J 1: 38–42

Treleaven J, Jull G, Atkinson L 1994 Cervical musculoskeletal dysfunction in post-concussional headache. Cephalgia 14: 273–279

Triano J, Schultz A B 1987 Correlation of objective measure of trunk motion and muscle function with low-back disability ratings. Spine 12: 561

Walsh E G 1992 Muscles, masses and motion: the physiology of normality, hypotonicity, spasticity, and rigidity. MacKeith Press, Blackwell Scientific Publications, Oxford

Watson D H, Trott P H 1993 Cervical headache: an investigation of natural head posture and upper cervical flexor muscle performance. Cephalgia 13: 272–284

ACKNOWLEDGEMENT

The author would like to thank Professor Vladimir Janda and Dr Karel Lewit for their pioneering work, upon which much of this chapter is based.

6

MET and the treatment of joints

While Janda (1988) acknowledges that it is not known whether dysfunction of muscles causes joint dysfunction or vice versa, he points to the undoubted fact that they massively influence each other, and that it is possible that a major element in the benefits noted following joint manipulation derives from the effects such methods (high velocity thrust, mobilisation etc.) have on associated soft tissues.

Steiner (1994) has discussed the influence of muscles in disc and facet syndromes. He describes a possible sequence as follows:

- A strain involving body torsion, rapid stretch, loss of balance etc. produces a myotatic stretch reflex response in, for example, a part of the erector spinae.
- The muscles contract to protect excessive joint movement, and spasm may result if (for any of a range of reasons, see notes on facilitation in Ch. 2, p. 38) there is an exaggerated response and they fail to assume normal tone following the strain.
- This limits free movement of the attached vertebrae, approximates them and causes compression and bulging of the intervertebral discs and/or a forcing together of the articular facets.
- Bulging discs might encroach on a nerve root, producing disc syndrome symptoms.
- Articular facets, when forced together, produce pressure on the intra-articular fluid, pushing it against the confining facet capsule which becomes stretched and irritated.
- The sinuvertebral capsular nerves may therefore become irritated, provoking muscular guarding, initiating a self-perpetuating process of pain–spasm–pain.

Steiner continues, 'From a physiological standpoint, correction or cure of the disc or facet syndromes should be the reversal of the process that produced them, eliminating muscle spasm and restoring normal motion.'

He argues that before discectomy or facet rhizotomy is attempted, with the all too frequent 'failed disc syndrome surgery' outcome, attention to the soft tissues and articular separation to reduce the spasm should be tried, in order to allow the bulging disc to recede and/or the facets to resume normal motion.

Clearly, osseous manipulation often has a place in achieving this objective, but the evidence of clinical experience indicates that a soft-tissue approach which either relies largely on MET, or at least which incorporates MET as a major part of its methodology, is frequently likely to produce excellent results in at least some such cases, and research evidence of this is available (see Ch. 7).

PREPARING JOINTS FOR MANIPULATION USING MET

What, though, if high velocity thrust or mobilisation methods of joint manipulation is the appropriate method of choice in treatment of a restricted joint? How does MET fit into the picture?

Muscle energy methods are versatile, and while they certainly have applications which are aimed at normalising soft tissue structures, such as shortened or tense muscles, with no direct implications as to the joints associated with these, they can also be used to help to normalise joint mobility via their influence on the associated soft tissues, which may be the major obstacle to the restoration of free movement.

As we have seen in previous chapters, MET may be employed to relax tight, tense musculature, or even spasm, and can also help to reduce the fibrotic changes in chronic soft-tissue problems, and tone weakened structures which may be present in the antagonists of shortened soft tissues. MET may therefore be employed in a pre-manipulative mode. In this instance, the conventional manipulative procedure is prepared for as it would normally be, whether this involves leverage or a thrust technique. The operator could then, having adopted an appropriate position, made suitable manual contacts and prepared the tissues for the adjustment by taking out available slack (manipulative effort), ask the patient to 'push back' against this position. The operator will have engaged the barrier in this preparation for manipulation, and will have taken out the slack that was available in the soft tissues of the joint(s), in order to achieve this position.

When the patient is asked to firmly but painlessly resist or 'push back', against the operator's contact hands, this produces a patient-indirect (operator pushing to the resistance barrier while patient pushes away from it) isometric contraction, which would have the effect of contracting the shortened muscles associated with the restricted joint.

After holding this effort for several seconds (ideally with a held breath) both operator and patient would simultaneously release their efforts, in a slow, deliberate manner. This can be repeated several times, with the additional slack being taken out after appropriate relaxation by the patient. Having engaged and re-engaged the barrier a number of times, the operator would decide when adequate release of restraining tissues had taken place and would then make the adjustment as normal.

Laurie Hartman (1985) states that: 'If the patient is in the absolute optimum position for a particular thrust technique during one of these repetitions (of MET), the joint in question will be felt to release. Even if this has not occurred, when retesting the movement range there is often a considerable increase in range and quality of play.'

He suggests that the operator use the temporary rebound reflex relaxation in the muscles, which will have followed the isometric contraction, to perform the technique. This will allow successful completion of the adjustment with minimal force. This refractory period of relaxation lasts for quite a few seconds and is valuable in all cases, but especially where the patient is tense or resistant to a manipulative effort.

Joint mobilisation using MET

The emphasis of MET on soft tissues should not be taken to indicate that intra-articular causes of dysfunction are not acknowledged.

Indeed, Lewit (1985) addressed this controversy in an elegant study which demonstrated that some typical restriction patterns remain intact even when the patient is observed under narcosis with myorelaxants. He tries to direct attention to a balanced view when he states, 'The naive conception that movement restriction in passive mobility is necessarily due to articular lesion has to be abandoned. We know that taut muscles alone can limit passive movement and that articular lesions are regularly associated with increased muscular tension.'

He then goes on to point to the other alternatives, including the fact that many joint restrictions are not the result of soft tissue changes, using as examples those joints not under the control of muscular influences – tibiofibular, sacroiliac, acromioclavicular. He also points to the many instances where *joint play* is more restricted than normal *joint movement*; since joint play is a feature of joint mobility which is not subject to muscular control, the conclusion has to be made that there are indeed joint problems which have the soft tissues as a secondary factor in any general dysfunctional pattern of pain and/or restricted range of motion (blockage). He continues, 'This is not to belittle the role of the musculature in movement restriction, but it is important to re-establish the role of articulation, and even more to distinguish clinically between movement restriction caused by taut muscles and that due to blocked joints, or very often, to both.'

Fortunately MET is capable of offering assistance in normalisation of both forms of dysfunction.

Basic criteria for treating joint restriction

In treating joint restriction with MET Sandra Yates (1991) suggests the following simple criteria be maintained:

1. The joint should be positioned at its physio-logical barrier – specific in three planes if spinal segments are being considered, flexion–extension, sidebending, rotation.
2. The patient should be asked to statically contract muscles towards their freedom of motion i.e. away from the barrier(s) of restriction, as the operator resists totally any movement of the part. The contraction, Yates suggests, is held for about 3 seconds (many MET experts suggest longer – up to 10 seconds).
3. The patient is asked to relax for 2 seconds or so between the contraction efforts, at which time,
4. The operator re-engages the joint at its new motion barrier(s).

This process is repeated until free movement is achieved or until no further gain is apparent following a contraction.

Precise focus of forces: example of lumbar dysfunction (Fig. 6.1)

Stiles (1984a), like most other practitioners using muscle energy methods, stresses the importance of accurate, precise, structural diagnosis if MET is to be used effectively in treatment of joint dysfunction. By careful motion palpation, determination is made as to restricted joints or areas, and which of their motions is limited. Precise, detailed localisation is required if there is to be accuracy in determining the direction in which the patient is to apply their forces, so that the specific restricted barrier can be engaged. If MET applications are poorly focused, it is possible to actually create hypermobility in neighbouring segments instead of normalising the restricted segment, by inappropriately introducing stretch into already adequately mobile tissues, above or below the restricted area.

For example, if a particular restriction is present in lumbar vertebrae, say limitation in gapping of the L4–L5, left side facets, on flexion; should a general MET mobilisation attempt, not localised to this segment, be used, which involved the joints above and/or below the restricted segment, hypermobility of these joints

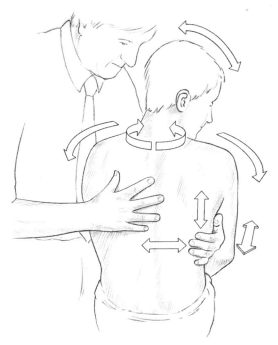

Figure 6.1A General assessment for restriction in thoracic spine, showing the possible directions of movement – flexion, extension, sidebending and rotation right and left, translation forwards, backwards, laterally in both directions, compression and distraction. MET treatment can be applied from any of the restriction barriers (or any combination of barriers) elicited in this way, with the area stabilised at the point of restriction.

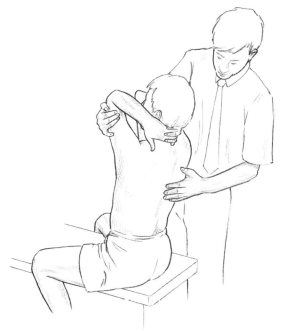

Figure 6.1C Assessment and possible MET treatment position for flexion restriction (inability to adequately extend) in the mid-thoracic area. MET treatment should commence from the perceived restriction barrier.

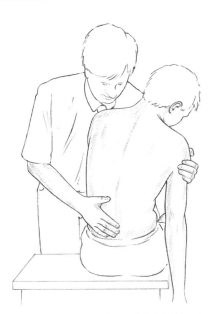

Figure 6.1B Assessment and possible MET treatment position for restriction in sidebending and rotation to the right, involving the lumbar spine.

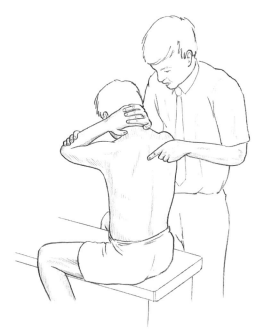

Figure 6.1D Assessment and possible MET treatment position for extension restriction (inability to adequately flex) in the upper thoracic area. MET treatment should commence from the perceived restriction barrier.

Figure 6.1E   Assessment and possible MET treatment position for sidebending restriction (inability to adequately sidebend left) in the mid-thoracic area. MET treatment should commence from the perceived restriction barrier.

Figure 6.1F   Assessment and possible MET treatment position for rotation restriction (inability to adequately rotate right) in the upper thoracic area. MET treatment should commence from the perceived restriction barrier.

could result leading, on retesting for general mobility, to an incorrect assumption that the restriction had been cleared.

In order to localise the effort at this segment, the patient would require to be positioned, so as to precisely engage the barrier in that joint. For example, one hand could gently palpate the facets of L4–L5, whilst the seated patient was guided into a flexed and sidebent position, which brought the affected segment to its barrier of motion. At that point an instruction for the patient to attempt to return to an upright position would involve the agonists restraining the joint from movement to its normal barrier, whilst the operator's force would be restraining any movement at all. This isometric contraction, should ideally be maintained for 3 to 5 seconds, (Stile's timing) with no more than perhaps 20% of the patient's strength being employed in the effort (and synchronised to breathing, as mentioned above, p. 118). After this, when all efforts had ceased, the barrier would normally be found to have retreated, so that greater flexion and sidebending could be achieved, without

effort, before re-engaging the barrier. Repetition would continue several times, until the maximum degree of motion had been obtained.

The exact opposite method could also be employed, in which, having engaged the barrier, the patient attempted to move through it, whilst being restrained. This would bring into play reciprocal inhibition of the contracted muscles, which might be involved in, though not necessarily causing, the inability of the joint to gap normally.

By using the antagonists instead of the affected muscles, there would be a lesser likelihood of pain being produced, were this an acute problem.

Goodridge (1981) states, 'Monitoring of force is more important than intensity of force. Localisation depends on the operator's palpatory proprioceptive perception of movement (or resistance to movement) at or about the specific articulation.' He continues, 'Monitoring and confining forces to the muscle group, or level of somatic dysfunction involved, are important in achieving desirable changes. Poor results are most often due to improperly localised forces, usually too strong.'

Localisation of restrictions, and identification of muscular contractions and fibrotic changes, is a matter of careful palpation, a set of skills which require constant refinement and maintenance by virtue of use.

Identification of the particulars of each restriction is critical, and this can only be achieved via the development of the skills required to assess joint mechanics, combined with a sound anatomical grasp. Assessment, via motion palpation,

is called for. If forces are misdirected then results will be poor, and may exacerbate the problem. Localisation of the point of restriction, in joint problems, is the major determining factor of the success (or otherwise) of MET (as in all manipulation).

Harakal's cooperative isometric technique (Harakal 1975) (Fig. 6.2)

When there is a specific or general restriction, in a spinal articulation (for example):

Figure 6.2A Harakal's approach requires the dysfunctional area (mid-thoracic in this example, in which segments cannot easily sidebend right and rotate left) to be taken to a position just short of the assessed restriction barrier. This is termed a point of 'balanced tension' where, after resting for a matter of seconds, an isometric contraction is introduced as the patient attempts to return towards neutral (sitting upright) against the operator's resistance.

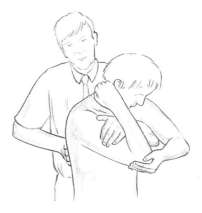

Figure 6.2B Following this effort, the restriction barrier should have eased and the patient can be guided through it towards a new point of balanced tension, just short of the new barrier, and the procedure is repeated.

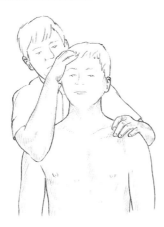

Figure 6.2C In this example the patient, who cannot easily sidebend and rotate the neck towards the left, is held just short of the present barrier in order to introduce an isometric contraction by turning the head to the right against resistance.

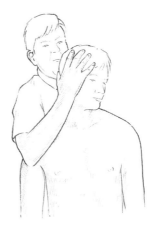

Figure 6.2D Following the contraction described in Figure 6.2C, it is possible for the operator to ease the neck into a greater degree of sidebending and rotation towards the left.

- The area should be placed in neutral (patient seated usually).
- The permitted range of motion should be determined by noting the patient's resistance to further motion.
- The patient should be rested for some seconds at a point just short of the resistance barrier, termed the 'point of balanced tension', in order to 'permit anatomic and physiologic response' to occur.
- The patient is asked to reverse the movement towards the barrier by 'turning back towards where we started' (thus contracting any agonists which may be influencing the restriction).
- The degree of patient participation at this stage can be at various levels, ranging from 'just think about turning' to 'turn as hard as you would like', or by giving specific instructions.
- Following a holding of this effort for a few seconds, and then relaxing completely, the patient is taken further in the direction of the previous barrier, to a new point of restriction determined by their resistance to further motion as well as tissue response (feel for 'bind').
- The procedure is repeated until no further gain is being achieved.
- It would also, of course, be appropriate to use the opposite direction of rotation, for example asking the patient to 'turn further towards the direction you are moving', so utilising the antagonists to the muscles which may be restricting free movement.

What if pain is produced when using MET in joint mobilisation?

Evjenth (Evjenth & Hamberg 1984) has a practical solution to the problem of pain being produced when an isometric contraction is employed. He suggests that the degree of effort be markedly reduced and the duration of the contraction increased, from 10 to up to 30 seconds.

If this fails to allow a painless contraction then use of the antagonist muscle(s) for the isometric contraction is another alternative.

Following the contraction, if a joint is being moved to a new resistance barrier and this produces pain, what variations are possible?

If, following an isometric contraction and movement towards the direction of restriction, there is pain, or if the patient fears pain, Evjenth suggests, 'then the therapist may be more passive and let the patient actively move the joint'.

Pain experienced may often be lessened considerably if the therapist applies gentle traction while the patient actively moves the joint. Sometimes pain may be further reduced if, in addition to applying gentle traction, the therapist simultaneously either aids the patient's movement at the joint, or provides gentle resistance while the patient moves the joint.

EXAMPLES OF MET IN JOINT TREATMENT

As we have seen, joints are treatable via MET, and some additional examples are given below. It is, however, not possible to provide a comprehensive body-wide, joint by joint, description of MET application in joint restriction, especially in a text focusing its attention on soft tissue dysfunction. Nevertheless, sufficient information is provided in this chapter to allow the interested therapist/practitioner to pursue this approach further, providing insights into possible technique applications involving spinal joints quite specifically, as well as generally, and also, more surprisingly perhaps, in dealing with joints which have no obvious muscular control, the iliosacral and acromioclavicular joints, both of which respond dramatically well to MET.

The low back

Diane Lee (Grieve 1986) describes a low back approach, using MET, which provides insights which can be adapted for use in other spinal regions. The example she gives is of a spine which is capable of full flexion, but in which palpable left sidebending, and left rotation fixation exists (i.e. it is locked in left sidebending rotation, and therefore cannot freely sidebend and rotate to the right) in the lumbar spine.

In Lee's example, the patient sits on a stool, feet apart and flat on the floor. The left arm hangs between the patient's knees, taking him into slight flexion and right rotation/side-bending. The operator stands at the patient's left side, with her left leg straddling the patient's left leg. The operator reaches across, and holds the patient's right shoulder, whilst the right hand palpates the vertebral interspace between the spinous processes immediately below the vertebra which is restricted in its ability to rotate to the right.

The patient is asked to slump forwards in this twisted posture, until the segment under inspection is most prominent, posteriorly. At this point the operator presses her left pectoral area against the patient's left shoulder and, with the patient still flexed, the spine is sidebent by the operator, without resistance, so that the patient's right hand approximates the floor. The operator then rotates the patient to the right, until maximum tension (bind) is felt to build at the segment being palpated. This is the barrier.

At this time the first MET procedure is brought into play. The patient is asked to attempt to reach the floor with his right hand, and this is resisted by the operator. This may last for 5 to 10 seconds, after which the patient relaxes (exhaling). The operator increases the sidebending and rotation to the right, before increasing the degree of flexion. No force is used, simply removal of whatever additional degree of slack has been produced by the isometric effort.

This is the new barrier, and the procedure of attempting to increase these directions of spinal movement (sidebending and rotation to the right, and flexion) is repeated, against resistance, a further 3 to 4 times. These movements all involve reciprocal inhibition of the shortened muscle fibres, which are holding the restricted area in left sidebending and left rotation. The antagonists are being contracted isometrically, to induce relaxation of the tense (agonist) structures.

After this, the patient, who is still flexed and rotated, and sidebent to the right, attempts to push against the operator's chest, with the left shoulder (i.e. attempts to rotate left and side-bend left, as well as to extend). This effort is maintained for a few (5 to 10) seconds before relaxation, re-engagement of the barrier, and repetition. This contraction involves those structures which have shortened, and so the isometric contraction produces postisometric relaxation in them. After each such contraction the slack is again taken out, by taking the patient further into right sidebending, rotation and flexion. The operator's position alters after the isometric efforts to the left and right, so that she now stands behind the patient with a hand on each shoulder.

Lee then suggests that the patient be asked to perform a series of stretching movements to the floor, first with the left hand and then with the right hand, against resistance, before being brought into an upright position by the operator, against slight resistance of the patient. The condition is then reassessed.

Notice that both reciprocal inhibition and postisometric relaxation are used in this manoeuvre. Lee (Grieve 1986) states 'Whether autogenic (PIR) or reciprocal inhibition is used is totally dependent on which technique effects the best neurophysiological change'.

In practice, however, it may not be clear which to choose; the author's experience is that PIR (what Lee chooses to call 'autogenic'), incorporating use of the agonists – those structures thought to be most restricted and negatively influencing joint movement – produces the most satisfying results. Reciprocal inhibition methods are, nevertheless, valuable – for example, in situations in which PIR is painful, or where agonist and antagonist are virtually interchangeable due to the nature of the problem (say, following trauma such as whiplash in which all soft tissues will have been stressed).

In short, PIR works best in chronic and RI in acute settings, but both can usefully be used in either type of condition if the guideline is adhered to that no pain should be produced, and if no attempt is made to force or 'stretch' joint structures.

Unlike the approach adopted in treating

muscles as such, joint applications of MET require that the barrier is all that is approached, with no attempt at pushing through it.

Additional choices

Goodridge (1981) describes two additional MET procedures to achieve the same end. The same pattern of dysfunction is assumed.
Goodridge states:

If the left transverse process of L5 is more posterior, when the patient is flexed, one postulates that the left caudad facet did not move anteriorly and superiorly along the left cephalad facet of S1, as did the right caudad facet. The movement to resolve the non-movement is postulated to have restrictions in minute movements in the directions of flexion, lateral flexion (side-bending) to the right and rotation to the right. It is further postulated, or conceptualised, that the non-moving side is restrained by hypertonicity (or shortening) of some muscle fibres. Therefore, the operator devises a muscle energy procedure to decrease the tone of (or to lengthen) the affected fibres.

The position, as described by Grieve above, is adopted. The patient is seated, left hand hanging between thighs, with the operator at his left, the patient's right hand lateral to his right hip and pointing to the floor. The operator's right hand monitors either L5 spinous or transverse process. The patient's left shoulder is contacted against the operator's left axillary fold, and upper chest. The operator's left hand is holding the patient's right shoulder. The patient slouches to flex the lumbar spine, so that the apex of the posterior convexity is located at the L5–S1 articulation. The operator induces first right sidebending, and then right rotation (patient's right hand approaches the floor) and localises movement at L5 when a sense of bind and restriction is noted there by the palpating hand. The patient is then asked to attempt to move in one or more directions, singly or in combination with each other. These would involve left sidebending, rotation left, and/or extension, all against operator's counterforces.

The patient is, in all of these efforts, contracting muscles on the left side of the spine, but is not changing the distance between the origin and insertion in muscles on either side of the spine. This achieves postisometric relaxation, and subsequent contractions would be initiated after appropriate taking up of slack and engagement of the new barrier. An alternative to this is that, having attained the position of flexion, right sidebending and right rotation localised at the joint in question, the patient is asked to move both shoulders in a translation to the left, against resistance from the operator's chest and left anterior axillary fold.

Neither of the shoulders should rise or fall as this is done, during the translation effort. Whilst the patient is attempting to move in this manner the operator palpates the degree of increased right sidebending which it induces, at L5–S1. As the patient eases off from this contraction, as described, the operator should be able to increase right rotation and sidebending until, once again, resistance is noted.

The objective of this alternative method is the same as in Grieve's example, but the movement involves, according to Goodridge, a concentric-isotonic procedure, because it allows right lateral flexion of the thoracolumbar spine during the effort.

As this demonstrates, some MET methods are very simple, while others involve conceptualisation of multiple movements, and the localisation of forces to achieve their ends. The principles remain the same, however, and can be applied to any muscle or joint dysfunction since the degree of effort, duration of effort and muscles utilised provide so many variables which can be tailored to meet most needs.

Cervical application of MET

Edward Stiles has described some of the most interesting applications of MET in treatment of joint restrictions. In this segment some of his thoughts on cervical assessment and treatment are explained (Stiles 1984b).

General procedure using MET for cervical restriction

Prior to any testing, Stiles suggests a general manoeuvre in which the patient is sitting upright. The operator stands behind and holds the head in the midline, with both hands stabilising it, and possibly employing her chest to prevent neck extension. The patient is told to try (gently) to flex, extend, rotate and sidebend the neck, in all directions, alternately. No particular sequence is necessary as long as all directions are engaged 5 or 6 times. Each muscle group should undergo slight contraction against unyielding force. This relaxes the tissues in a general manner. Traumatised muscles will relax without much pain, via this method.

Upper cervical dysfunction assessment and MET treatment

To test for dysfunction in the upper cervical region, the patient lies supine. The operator flexes the head on the neck slightly, with one hand, whilst the other cradles the neck. Flexion of this small degree stabilises the cervical area below C2, so that evaluation of atlanto-axial rotation may be carried out. The region C1 and C2 is usually responsible for half the gross rotation of the neck. With the neck flexed (effectively 'locking' everything below C2) it is then passively rotated to both left and right. If the range is greater on one side, then this is indicative of a probable restrictive barrier, which may be amenable to MET. If rotation towards the left is normally about 85°, but in this instance it is restricted, then palpation of muscle tissues at the level of the facets of C2 (just below the level of joint dysfunction) should indicate contraction or tension locally on the right. This may or may not be tender, but the likelihood is that it will be so if there is dysfunction. (Pain is often more noticeable at the level of any hypermobile joint rather than where the actual restriction is noted. This may be ascertained by palpation and motion palpation, feeling the tissues as the joint is moved).

If dysfunction is suspected at the atlanto-axial joint, then C2 is stabilised, in order to isolate C1 for treatment. A fingertip is placed on the left transverse process of C2, so that it cannot turn left when the patient's head is turned left. The second finger of the operator's left hand (which is cradling the neck in flexion) is rested as a

barrier to prevent left rotation of C2, and the head is then taken gently into left rotation. C1 and the head move, and C2 remains fixed. The barrier is engaged when C2 starts to move, i.e. 'bind' is noted by the palpating finger.

The slack is removed, and at that point the patient is asked to try to turn the head gently to the right, away from the barrier. The operator's right hand should be resting on the right side of the patient's head, to prevent this right rotation. The patient's light rotation force is exerted against the operator's hand and this is maintained for a few seconds (4 to 10). Both patient and operator release their efforts at the same time, and the operator then attempts to take the head further to the left, without force, to engage the new barrier.

This is repeated 2 or 3 times. This monitoring and stabilising pressure on C2 is minimal, since the patient's effort is not a strong one (this must be stressed to the patient). The patient is using the muscles which are in spasm, or contracted (preventing rotation left) and, according to Stiles, 'the exertion builds up tension in the contracted muscles; the Golgi receptor system starts reporting the increased tension in relation to surrounding muscles, and spasm is reflexively inhibited.' This is an operator-direct approach, involving postisometric relaxation.

Stiles's comments regarding whiplash injury and MET

In such conditions X-ray pictures are often normal, as are neurological examinations. Pain, often of major proportions, is nevertheless present. Careful examination should show some segments which are not capable of achieving a full range of movement. These would normally correlate with palpable tissue change and sensitivity. More often than not there is a restriction in which a vertebra is caught in flexion (forward bending). Less commonly, extension fixations may be noted. Each vertebra should be tested to note its ability to flex, extend, sidebend and rotate. MET is applied to whatever specific restrictions are found, as in the example above (p. 120).

Wherever a restriction is noted, in any

particular direction, MET should be used. For example, if C3–4 facets close properly as the neck is sidebent to the left, a characteristic physiological 'springing' will be noted as the barrier is reached. If on the right, however, there is dysfunction as the neck is sidebent to that side, the facets will not be felt to close, and a pathological barrier will be noted, which is characterised by a lack of 'give', or increased resistance. This restriction may be expressed in two ways:

- The positional diagnosis would be that the segment is flexed and sidebent to the left (and therefore, because of the nature of spinal mechanics, rotated left).
- The functional diagnosis would be that the joint will not extend, sidebend, or rotate to the right.

The patient should be in the same position as was used in diagnosis (supine, neck slightly flexed). The operator's right middle fingers would be placed over the right pillars of C3–4, and the neck taken to the maximum position of sidebending rotation to the right, engaging the barrier.

The left hand is placed over the patient's left parietal and temporal areas. With this hand offering counterforce, the patient is invited to sidebend and rotate to the left, for a few seconds. This employs the muscles which are shortened, and which are preventing the joint from easily sidebending and rotating to the right. Postisometric relaxation of these will follow, and the neck should be taken to its new barrier, and the same procedure repeated 2 or 3 times. An alternative would be for the patient to engage the barrier whilst the operator resisted, so incorporating reciprocal inhibition.

Greenman's exercise in cervical palpation and MET use

The following exercise sequence is based on the work of Philip Greenman, and is suggested as an excellent way of becoming familiar with both the mechanics of the neck joints, and safe and effective MET applications to whatever is found to be restricted (Greenman 1989).

In performing this exercise it is important to be aware that normal physiology dictates that sidebending and rotation in the cervical area is 'type 2', which means that segments which are sidebending will automatically move towards the same side. A sidebend to the right means that rotation will take place to the right. Most cervical restrictions are compensations and will involve several segments, all of which will adopt this 'type 2' pattern. Exceptions occur if a segment is traumatically induced into a different format of dysfunction, in which case there would be sidebending to one side and rotation to the other – termed 'type 1', which is the physiological pattern for the rest of the spine.

To easily palpate for sidebending and rotation, a side to side translation movement is used, with the neck in slight flexion or slight extension. When the neck is absolutely neutral (no flexion or extension – an unusual state in the neck) true translation side to side is possible. As a segment is translated to one side it is therefore automatically sidebending and, because of the anatomical and physiological rules governing it, it will be rotating to the same side.

In order to evaluate cervical function using this knowledge, Greenman suggests that the operator places the fingers as follows, on each side of the spine (see Fig. 6.3):

- The index finger pads rest on the articular pillars of C6, just above the transverse processes of C7, which can be palpated just anterior to the upper trapezius
- The middle-finger pads will be on C6 and the ring fingers on C5, with the little finger pads on C3.

Then:

1. With these contacts (operator seated at the head of the supine patient) it is possible to examine for sensitivity, fibrosis, hypertonicity as well as being able to apply lateral translation to cervical segments with the head in flexion or extension. In order to do this effectively, it is necessary to stabilise the superior segment to the one about to be examined. The heel of the hand controls movement of the head.

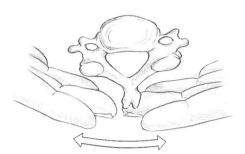

Figure 6.3A The finger pads rest as close to the articular pillars as possible, in order to be able to palpate and guide vertebral motion in a translatory manner.

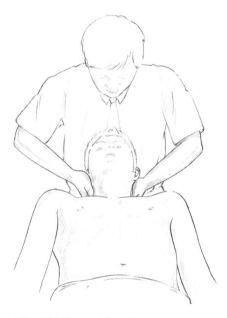

Figure 6.3B With the neck/head in a neutral position, the operator sequentially guides individual segments into translation in both directions in order to sense indications of restriction and tissue modification. If a restriction is sensed, its increase or decrease is evaluated by retesting with the segment held in greater flexion and then extension. MET would be applied from the position of greatest unforced bind/restriction, using muscles which would either take the area through (antagonists to shortened muscles) or away from the barrier (shortened muscles themselves – the agonists).

2. With the head/neck in relative neutral (no flexion and no extension) translation to the right and then left is introduced (any segment) to assess freedom of movement (sidebending and rotation) in each direction.

Say C5 is being stabilised with the fingerpads, as translation to the left is introduced.

The ability of C5 to freely sidebend and rotate on C6 is being evaluated with the neck in neutral.

If the joint is normal, this translation will cause a gapping of the left facet and a 'closing' of the right facet as left translation is performed, and vice versa. There will be a soft end-feel to the movement, without harsh or sudden braking.

If, say, translation of the segment towards the right from the left produces a sense of resistance/bind, then the segment is restricted in its ability to sidebend left and (by implication) to rotate left.

3. If such a restriction is noted, the translation should be repeated, but this time with the head in extension instead of neutral. This is achieved by lifting the contact fingers on C5 (in this example) slightly towards the ceiling before reassessing the side to side translation.

4. The head and neck are then taken into flexion, and left to right translation is again assessed.

The objective is to ascertain which position creates the greatest degree of bind as the barrier is engaged. Is it more restricted in neutral, extension, flexion?

If this restriction is greater with the head extended, the diagnosis is of a joint locked in flexion, sidebent right and rotated right (meaning that there is difficulty in the joint extending and of sidebending and rotating to the left).

If this (C5 on C6 translation left to right) restriction is greater with the head flexed, then the joint is locked in extension and sidebent right and rotated right (meaning there is difficulty in the joint flexing, sidebending and rotating to the left).

MET treatment of the cervical area

Using MET, and using the same example (C5 on C6 as above, translation left is restricted with the greatest degree of restriction noted in extension) the procedure would be as follows.

One hand palpates both of the articular pillars of the inferior segment of the pair which is dysfunctional.

In this instance, this hand will stabilise the C6 articular pillars, holding the inferior vertebra so that the superior segment can be moved on it.

The other hand will introduce movement to, and control the head and neck above, the restricted vertebra.

The articular pillars of C6 are held and are lifted towards the ceiling, introducing extension, while the other hand introduces sidebending and rotation to the right until the restriction barrier is reached.

A slight isometric contraction is introduced using sidebending, rotation or flexion (or all of these). The patient is asked to try to lightly turn his head to the left and to sidebend it that way while straightening the neck, or any one of these movements, which should be firmly restrained.

After 5 to 7 seconds the patient relaxes, and extension, sidebending and rotation right are increased to the new resistance barrier, with no force at all. Repeat 3 or 4 times.

Eye movement can be used instead of muscular effort in cases where effort results in pain. Looking upwards will encourage isometric contraction of the extensors and vice versa, and looking towards a direction encourages contraction of the muscles on that side.

MET treatment of acromioclavicular and sternoclavicular dysfunction

Whereas spinal/neck and most other joints are seen to be moved by and to be under the postural influence of muscles, and therefore to an extent to be capable of having their function modified by muscle energy techniques, articulations such as those of the sternoclavicular, acromioclavicular and iliosacral joints seem far less amenable to such influences. Hopefully, some of the methods detailed below will modify this impression, since MET is widely used in the osteopathic profession to help normalise the functional integrity of these joints.

Acromioclavicular (AC) dysfunction

Stiles suggests beginning evaluation of AC dysfunction at the scapula, the mechanics of which closely relate to AC function.

The patient sits erect and the spines of both scapulae are palpated by the operator, standing behind. The hands are moved medially, until the medial borders of the scapulae are identified, at the level of the spine. Using the palpating fingers as landmarks, the levels are checked to see whether they are the same. Inequality suggests AC dysfunction. The side of dysfunction remains to be assessed and each is tested separately (see Fig. 6.4). To test the right side AC joint, the operator is behind the patient, with the left hand palpating over the joint. The right hand holds the patient's right elbow. The arm is lifted in a direction, 45° from the sagittal and frontal planes. As the arm approaches 90° elevation, the AC joint should be carefully palpated for hinge movement, between the acromion and the clavicle.

In normal movement, with no restriction, the palpating hand should move slightly caudad, as the arm is abducted beyond 90°.

If the AC is restricted the palpating hand/ digit will move cephalad, and little or no action will be noted at the joint itself, as the arm goes beyond 90° elevation.

Muscle energy technique is employed with the arm held at the restriction barrier, as for testing above.

If the scapula on the side of dysfunction had been shown to be more proximal than that on the normal side, then the humerus is placed in external rotation, which takes the scapula caudad against the barrier, before the isometric contraction commences.

If, however, the scapula on the side of the AC dysfunction was more distal than the scapula on the normal side, then the arm is internally rotated, taking the scapula cephalad against the barrier before the isometric contraction commences.

The left hand (we assume this to be a right-sided problem in this example) stabilises the distal end of the clavicle, with caudad pressure being applied by the left thumb which rests on the proximal surface of the scapula. The first finger of the left hand lies on the distal aspect of the clavicle. The combination of the rotation of the arm as appropriate (externally if the scapula on that side was high and internally if it was low) as well as the caudad pressure exerted by

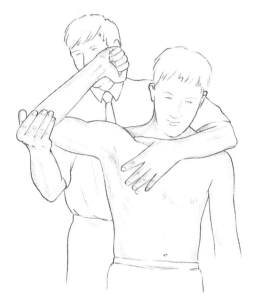

Figure 6.4A   MET treatment of right side acromioclavicular restriction. Patient attempts to return the elbow to the side against resistance.

Figure 6.4B   Following the isometric contraction, the arm is elevated further while firm downward pressure is maintained on the lateral aspect of the clavicle.

the left hand on the clavicle and the scapula, provides an unyielding counterforce. The arm will have been raised until the first sign of inappropriate movement at the AC joint was noted (as a sense of 'bind'). This is the barrier,

and at this point the various stabilising holds (internal or external arm rotation, etc.) are introduced. An unyielding counterpressure is applied at the point of the patient's elbow by the right hand, and the patient is asked to try to take that elbow towards the floor with less than full strength. After 7 to 10 seconds the patient and operator relax, and the arm is once more taken towards the barrier. Again, greater internal or external rotation is introduced, to take the scapula higher or lower, as appropriate, as firm but not forceful pressure is sustained on the clavicle and scapula in a caudad direction.

The mild isometric contraction is again called for, and the procedure repeated several times. It is worth recalling that respiratory accompaniment to the efforts described is helpful, with inhalation accompanying effort, and exhalation accompanying relaxation and the engagement of the new barrier. The procedure is repeated until no further improvement is noted in terms of range of motion or until it is sensed that the clavicle has resumed normal function.

Assessment and MET treatment of restricted abduction in the sternoclavicular joint

As the clavicle abducts it rotates posteriorly. To test for this motion, the patient lies supine, or is seated, with arms at side (Fig. 6.5A). Operator places index fingers on superior surface of medial end of the clavicle. The patient is asked to shrug the shoulders as the operator palpates for the expected caudal movement of the medial clavicle. If it fails to do so there is a restriction preventing normal abduction.

MET treatment. Operator stands behind seated patient with thenar eminence on the superior margin of the medial end of the clavicle to be treated. The other hand grasps the patient's flexed elbow and holds this at 90°, with the upper arm externally rotated and abducted (Fig. 6.5B). The patient is asked to adduct the upper arm for 5 to 7 seconds against resistance using about 20% of available strength. Following the effort and complete relaxation, the arm is abducted further, and externally rotated further, until a new barrier is sensed, all the while maintaining firm caudad pressure on the medial end

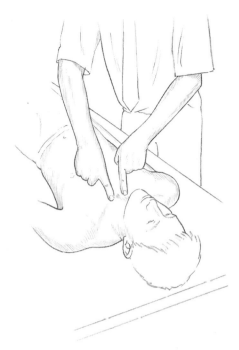

Figure 6.5A Assessment ('shrug test') for restriction in clavicular mobility.

Figure 6.5B MET treatment of restricted sternoclavicular joint. Following an isometric contraction, the arm is elevated and extended while firm downward pressure is maintained on the medial aspect of the clavicle with the thenar eminence.

of the clavicle. The process is repeated until free movement of the medial clavicle is achieved.

Assessment and MET treatment of restricted horizontal flexion of the upper arm (sternoclavicular restriction)

The patient lies supine and operator stands to one side with the index fingers resting on the anteromedial aspect of each clavicle.

The patient is asked to extend his arms forwards in front of his face, in a 'prayer' position, palms together, pointing to the ceiling (Fig. 6.6A). On pushing the hands forwards towards the ceiling, the clavicular heads should drop towards the floor, and not rise up to follow the hands. If one or both fail to drop, there is a restriction.

MET treatment involves the placement of a thenar eminence over the medial end of the clavicle, holding it towards the floor. The other hand lies under the shoulder on that side to embrace the dorsal aspect of the scapula (Fig. 6.6B).

The patient is asked to stretch out the arm on the side to be treated so that the hand can rest behind the operator's neck or shoulder. The operator leans back to take out all the slack from the extended arm and shoulder while at the same time lifting the scapula on that side slightly from the table. At this time the patient is asked to pull the operator towards him, against firm resistance for 7 to 10 seconds. Following complete release of all the patient's efforts, the downwards thenar eminence pressure – to the floor – is maintained (painlessly) and more slack is taken out (operator keeps in place all elements of the procedure throughout, only the patient releases effort between contractions). The process is repeated once or twice more or until the 'prayer' test proves negative. No pain should be noted during this procedure.

Assessment and MET treatment of iliosacral restrictions

In order to apply certain useful MET applications to pelvic dysfunction involving the iliosacral joint it is necessary to carry out a few basic assessments.

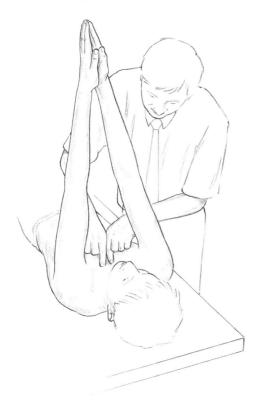

Figure 6.6A Assessment ('prayer test') for restricted horizontal flexion of the sternoclavicular joint.

Figure 6.6B MET treatment of horizontal flexion restriction. After isometric contraction (patient attempts to pull operator towards himself) the operator simultaneously lifts the shoulder while maintaining firm downwards pressure (to the floor) with the hypothenar eminence on the medial aspect of the clavicle.

CAUTION: The evidence derived from the standing flexion test as described below is invalid if there is concurrent shortness in the hamstrings, since this will effectively give either:

- A false positive sign on the contralateral side with unilateral hamstring shortness due to the restraining influence on the side of hamstring shortness, creating a compensating iliac movement on the other side during flexion, or
- False negative signs if there is bilateral hamstring shortness (i.e. there may be iliosacral motion which is masked by the restriction placed on the ilia via hamstring shortness).

The hamstring tests as described in Chapter 4 (p. 76) should therefore be carried out first, and if this proves positive these structures should be normalised prior to the assessment methods described here being utilised.

Test (a) Operator squats behind standing patient and places the medial side of their hands on the lateral pelvis below the crests pushing inwards and upwards until the index fingers lie superior to the crest (Fig. 6.7A). If these are judged to be level then no anatomical leg length discrepancy exists.

Test (b) The PSIS positions are assessed just below the pelvic dimples.
—Are they symmetrical?
—Is one superior or anterior to the other?

Anteriority may involve shortness of the external rotators on that side (iliopsoas, quadratus femoris, piriformis) or internal rotators on the other side (gluteus medius, hamstrings).

Inferiority may indicate hamstring shortness or pelvic/pubic dysfunction.

Test (c) With the patient still standing and any inequality of leg length having been compensated for by insertion of a pad under the foot on the short side, the thumbs are placed firmly (a light contact is useless) on the inferior slope of the PSIS.

The patient is asked to go into full flexion while thumb contact is maintained, with the operator's eyes at the level of the thumbs. The patient's knees should remain extended during this bend.

Observe, especially near the end of the excursion of the bend, whether one or other thumb seems to start to travel with the PSIS.

Interpretation: If one thumb moves superiorly during flexion it indicates that the ilium is fixed to the sacrum.

Test (d) Before performing the subsequent test, the seated flexion test, move to the front of the patient and look down the spine for evidence of greater fullness on one side or the other of the lumbar spine, indicating muscular mounding, possibly in association with spinal rotoscoliosis (or due to excessive tension in quadratus lumborum and/or the erector spinae).

Test (e) The seated flexion test involves exactly the same hand placement and observation of the thumb movement, if any, during full flexion, while the patient is seated on the table, legs over the side, knees in flexion (Fig. 6.7B).

Interpretation: In this instance, since the ischial tuberosities are being 'sat upon', the ilia cannot easily move and if one thumb travels forward during flexion it means that the sacrum is stuck to the ilium on that side, dragging the ilium with it in flexion.

Test (f) With the seated patient still fully flexed, move to the front and look down the spine for fullness in the paravertebral muscles, in the lumbar area. If greater fullness exists in a paraspinal area of the lumbar spine, with the patient standing as opposed to seated, then this is evidence of a compensatory process, involving the postural muscles of the lower limbs and pelvic area, as a prime cause.

If, however, fullness in the lumbar paraspinal region is the same when seated, or greater when seated, this indicates some primary spinal

Figure 6.7A Standing flexion test for iliosacral restriction. The dysfunctional side is that on which the thumb moves during flexion.

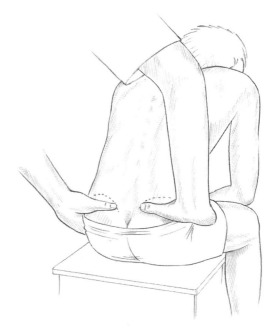

Figure 6.7B Seated flexion test for sacroiliac restriction. The dysfunctional side is that on which the thumb moves during flexion.

dysfunction and not a compensation for postural muscle imbalances.

Test (g) Confirmation of iliosacral restriction test: The patient stands, operator is behind, kneeling with thumbs placed so that on the side being assessed the contact is on the PSIS, while the other palpates the median sacral crest directly parallel to the PSIS.

The patient is asked to slowly and fully flex the ipsilateral hip to waist level.

If on the flexing of the hip there is a movement of the PSIS and the median sacral crest 'as a unit', together with a compensating adaptation in the lumbar spine, this indicates iliosacral restriction on that side.

If this combined movement occurred when the contralateral hip was being flexed it would indicate a sacroiliac restriction on the side being palpated.

What type of iliosacral dysfunction exists?

Once an iliosacral restriction has been identified, it is necessary to define precisely what type of restriction this is. The choices are limited to anterior rotation, posterior rotation, inflare and outflare, and this part of the evaluation process depends upon observation of landmarks.

Test (h) The patient lies supine and straight while the operator locates the inferior slopes of the two ASISs with thumbs, and views these contacts from directly above the pelvis with the dominant eye over centre line (bird's eye view – see Fig. 6.8A).
— Which thumb is nearer the head and which nearer the feet?
— Is one side superior or is the other inferior?
In other words, has one ilium rotated posteriorly or the other anteriorly?

This is determined by referring back to the standing flexion test (p. 133). The side of dysfunction as determined by the standing flexion test 'travelling thumb' and/or the standing hip flexion test is the side which is restricted, and therefore defines which landmark is taken into consideration (Fig. 6.8B).

If the thumb on one side travelled with the ilium on flexion, even if this was near the end of the range, or a positive hip flexion test, then this is the dysfunctional side, and if that is the side which appears inferior (compared to its pair) it is assumed that the ilium on the inferior side has rotated anteriorly on the sacrum.

If, on the side of the positive flexion test or hip flexion test, the ASIS appears superior to its pair, then the ilium on the superior side has rotated posteriorly on the sacrum.

While in the same position, to observe the ASIS positions, note is made as to the relative positions of these landmarks in relation to the midline of the patient's abdomen, using either the linea alba or the umbilicus as a guide.

The operator's eyes should be directly over the pelvis and the thumbs rest on the ASISs.
— Is one thumb closer to the umbilicus than the other? It is necessary at this stage to once again refer to which side is dysfunctional.
— Is the ASIS on the side which is further from the umbilicus outflared or is the ASIS which is closer to the umbilicus indicative of that side being inflared?

The ASIS associated with the side on which the thumb travelled is the dysfunctional side and the decision as to whether it is an inflare (closer to umbilicus) or outflare (further from umbilicus) is therefore obvious.

Flare dysfunctions are usually treated prior to rotation dysfunctions.

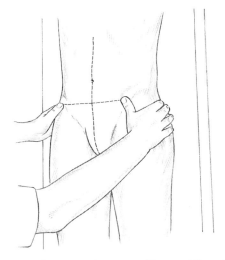

Figure 6.8A Operator adopts a position providing a bird's-eye view of ASIS prominences on which rest the thumbs.

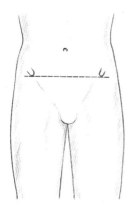

Figure 6.8Bi)   The ASISs are level and there is no rotational dysfunction involving the iliosacral joints.

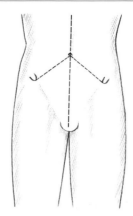

Figure 6.8Biii)   The ASISs are equidistant from the umbilicus and the midline, and there is no iliosacral flare dysfunction.

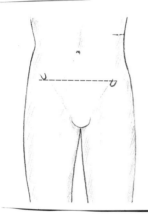

Figure 6.8Bii)   The right ASIS is higher than the left ASIS. If a thumb 'travelled' on the right side during the standing flexion test this would represent a posterior right iliosacral rotation dysfunction. If a thumb 'travelled' on the left side during the test this would represent an anterior left iliosacral rotation dysfunction.

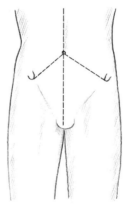

Figure 6.8Biv)   The ASIS on the right is closer to the umbilicus/midline which indicates that either there is a right side iliosacral inflare (if the right thumb moved during the standing flexion test), or there is a left side iliosacral outflare (if the left thumb moved during the standing flexion test).

MET treatment of iliac inflare (Fig. 6.9)

The patient is supine and the operator stands on the same side as the problem, with the cephalad hand stabilising the non-affected side ASIS and the caudad hand holding the ankle of the affected side (Fig. 6.9A).

The affected side hip is flexed and abducted while full external rotation is introduced to the hip. The operator's forearm aligns with the lower leg, elbow stabilising the medial aspect of the knee.

The patient is asked to lightly adduct the hip against the resistance offered by the restraining arm for 10 seconds while holding the breath. On complete relaxation and on an exhalation, with the pelvis held stable by the cephalad hand, the flexed leg is taken into more abduction and external rotation, if new 'slack' is now available.

This process is repeated once or twice, at which time the leg is slowly straightened while abduction and external rotation of the hip are maintained. The leg is then returned to the table.

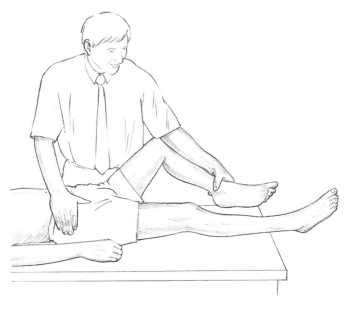

Figure 6.9A   An MET treatment position for left side iliosacral inflare dysfunction. Note the stabilising hand on the right ASIS.

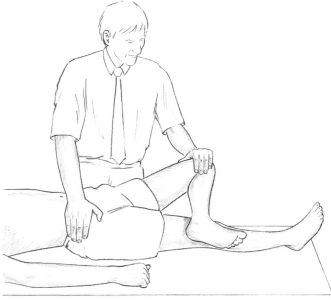

Figure 6.9B   An alternative MET treatment position for left side iliosacral inflare dysfunction. Note the stabilising hand on the right ASIS.

Care should be taken not to use the powerful leverage available from the flexed and abducted leg; its own weight and gravity provide adequate leverage and the 'release' of tone achieved via isometric contractions will do the rest. It is very easy to turn an inflare into an outflare by overenthusiastic use of force. The degree of flare should be re-evaluated and any rotation then treated (see below, pp 137–138).

MET treatment of iliac outflare (Fig. 6.10)

The patient is supine and the operator is on the same side as the dysfunctional ilium, supinated cephalad hand under the patient's buttocks with finger tips hooked into the sacral sulcus on the same side.

The caudad hand holds the patient's foot on the treated side, with the forearm resting along

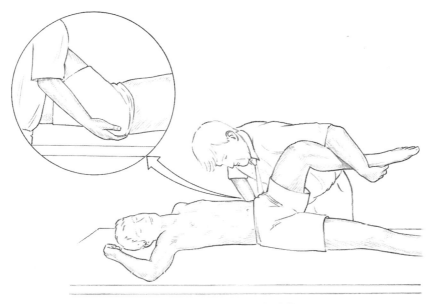

Figure 6.10   MET treatment of iliosacral outflare on the left.

the medial calf/shin area as the hand grasps the foot.

The hip on the treated side is fully flexed and adducted and internally rotated, at which time the patient is asked to abduct the hip against resistance using up to 50% of strength, for 10 seconds while holding the breath.

Following this and complete relaxation, slack is taken out and the exercise repeated once or twice more. As the leg is taken into greater adduction and internal rotation, to take advantage of the release of tone following the isometric contraction the fingers in the sacral sulcus exert a traction towards the operator, effectively guiding the ilium into a more inflared position.

After the final contraction, adduction and internal rotation are maintained as the leg is slowly returned to the table.

The evaluation for flare dysfunction is then repeated and if relative normality has been restored any rotational dysfunction is then treated, as per the methods described below.

MET treatment of anterior iliac rotation (Fig. 6.11)

The patient is prone. The operator stands at the side to be treated, at waist level. The affected leg and hip are flexed and brought over the edge of the table. The foot/ankle area is grasped between the operator's legs. The tableside hand stabilises the sacral area while the other hand

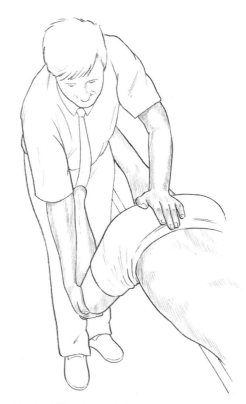

Figure 6.11   MET treatment of an anterior iliosacral restriction.

supports the flexed knee and guides it into greater flexion, inducing posterior iliac rotation, until the restriction barrier is sensed:

- By the palpating 'sacral contact' hand
- By virtue of a sense of greater effort in guiding the flexed leg
- By observation of pelvic movement as the barrier of resistance is passed.

The patient is asked to inhale, to hold the breath and to attempt to straighten the leg against unyielding resistance, for 10 seconds using no more than 20% of available strength.

On releasing the breath and the effort, and on complete relaxation and on an exhalation, the leg/innominate is guided to its new barrier.

Subsequent contractions can involve different directions of effort ('try to push your knee sideways', or ' try to bend your knee towards your shoulder' etc.) in order to bring into operation a variety of muscular factors to encourage release of the joint.[1]

The standing flexion test (p. 133) should be performed again to establish whether the joint is now free.

MET for treatment of posterior iliac rotation (Fig. 6.12)

The patient is prone and the operator stands on the side opposite the dysfunctional iliosacral joint. The tableside hand supports the anterior aspect of the patient's knee while the other rests on the PSIS of the affected side to evaluate bind.

The affected leg is hyperextended until free movement ceases, as evidenced by the following observations:

- Bind is noted under the palpating hand
- Sacral and pelvic motion are observed as the barrier is passed
- A sense of effort is increased in the arm extending the leg.

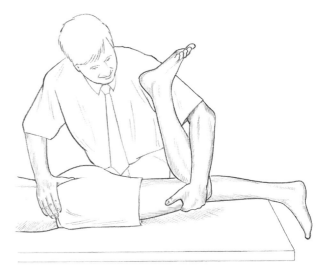

Figure 6.12   MET treatment of a posterior iliosacral restriction.

Holding the joint at its restriction barrier, the patient is asked, with no more than 20% of strength, to flex the hip against resistance for 10 seconds while holding the breath. After cessation of the effort, releasing the breath, and completely relaxing, on an exhalation, the leg is extended further to its new barrier.

No force is used at all; the movement after the contraction simply takes advantage of whatever slack is now available.

Variations in the direction of the contraction are sometimes useful if no appreciable gain is achieved using hip and knee flexion; perhaps involving abduction or adduction or even attempted extension.

The standing flexion test (p. 133) is performed again to establish whether iliosacral movement is now free, once a sense of 'release' has been noted following one of the contractions.

MET treatment for temperomandibular joint dysfunction

Dysfunction of the TMJ is a vast subject, and the implications of such problems have been related to a variety of other areas of dysfunction, ranging from cranial lesions to spinal and general somatic alterations and endocrine imbalance (Gelb 1977). The reader is referred to

---

[1]The same mechanics precisely can be incorporated into a side-lying position. The only disadvantage of this is the relative instability of the pelvic region compared to that achieved in the prone position described above.

Janda's observations on postural influences on TMJ problems (Ch. 2, pp 33–34).

Diagnosis of the particular pattern of dysfunction is, of course, essential before safe therapeutic intervention is possible. There are many possible causes of TMJ dysfunction, and a cooperative relationship with a skilled dentist is an advantage in such problems, since many aspects relate to the presence of faults in the bite of the patient.

A knowledge of cranial mechanics is useful, and a history of trauma should be sought in those patients presenting with TMJ involvement. One common source of injury is the equipment used in applying spinal traction, in which a head halter with a chinstrap is used. This can cause the mandible to be forced into the fossae, impacting the temporal bones into internal rotation. A strap causing pressure on the occipital region could jam the occipito-mastoid and lambdoid sutures, upwards and forwards, also resulting in internal rotation of the temporals. This can cause major dysfunction of cranial articulation and function which would be further exaggerated were there imbalances present in these structures, prior to the trauma. Inept manipulative measures can also traumatise the area, especially thrusting forces exerted onto the occiput whilst the head and neck are in extreme rotation. Any situation in which the patient is required to maintain the mouth opened for lengthy periods, such as during dental work, or when a laryngoscope is being used, may induce strain, especially if the neck is extended at the time. All, or any, such patterns of injury should be sought when TMJ pain, or limitation of mouth opening, is observed. Apart from correction of cranial dysfunction via skilled cranial osteopathic work, the muscular component invites attention, using MET methods and other appropriate measures. Gelb suggests a form of MET which he terms 'stretch against resistance' exercises.

MET method 1 Reciprocal inhibition is the objective when the patient is asked to open the mouth against resistance applied by the operator's, or the patient's, own hand (Fig. 6.13A). (Patient places elbow on table, chin in hand and attempts to open mouth against resistance for 10 seconds or so.) The jaw would have been opened to its comfortable limit before attempting this, and after the attempt it would be taken to its new barrier before repeating. This MET method would have a relaxing effect on the muscles which are shortened, or tight.

MET method 2 To relax the short tight muscles, using postisometric relaxation, counterpressure would be required in order to prevent the open jaw from closing (using minimal force). This would require the thumbs (suitably protected) to be placed along the superior surface of the lower back teeth, whilst an isometric contraction was performed by the patient (Fig. 6.13B). In this exercise the operator is directing force through the barrier, (operator direct method) rather than the patient (patient direct) as in first example.

MET method 3 Lewit suggests the following method of treating TMJ problems, maintaining that laterolateral movements are important, using postisometric relaxation. The patient sits with the head turned to one side (say the left in this example). The operator stands behind him and stabilises the patient's head against his chest (Fig. 6.13C). The patient opens his mouth, allowing the chin to drop, and the operator

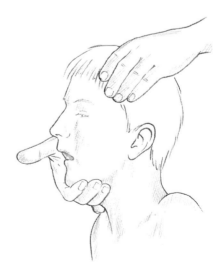

Figure 6.13A MET treatment of TMJ restriction, involving limited ability to open the mouth. The isometric contraction phase of treatment is illustrated as the patient attempts to open against resistance.

Figure 6.13B   MET treatment of TMJ restriction, involving an isometric contraction in which the patient attempts to close the mouth against resistance.

Following both these procedures, the patient would be encouraged to gently stretch the muscles by opening the mouth widely. This can be assisted by the operator.

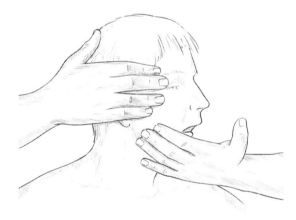

Figure 6.13C   MET treatment of lateral restrictions of the TMJ. Following the isometric contraction as described, the lateral excursion is increased.

cradles the mandible with her left hand, so that the fingers are curled under the jaw, away from her. The operator draws the jaw gently towards her chest, and when the slack has been taken up, the patient offers a degree of resistance to its being taken further, laterally. After a few seconds of gentle isometric contraction, the operator and patient relax simultaneously, and the jaw will usually have an increased lateral excursion. This is repeated 3 times. This should be performed so that the lateral pull is away from the side to which the jaw deviates, on opening.

Self-treatment exercise 1 Gelb suggests a retrusive exercise be used in conjunction with the above, both methods being useful in eliminating 'clicks' on opening the mouth. The patient curls the tongue upwards, placing the tip as far back on the roof of the mouth as possible. Whilst this is maintained in position, the patient is asked to slowly open and close the mouth (gently), to reactivate the suprahyoid, posterior temporalis and posterior digastric muscles (the retrusive group).

Self-treatment exercise 2 The patient places an elbow on a table, jaw resting on the clenched fist. This offers some resistance to the slow opening of the mouth. This is done 5 times with hand pressure, and then 5 times without, ensuring that the lower jaw does not come forward. The lower teeth should always remain behind the upper teeth on closing. A total of 25 such movements are performed, morning and evening.

Joints, end-feel and MET

In the text thus far there have been descriptions of MET applications to spinal, cervical, pelvic, acromioclavicular and sternoclavicular joints and TMJ, as well as to a multitude of muscles relating to almost all the joints in the body.

In order to apply the principles embodied in MET methodology to joint dysfunction not specifically covered in the text, all that is required is an appreciation of restriction barriers, which in essence means having an awareness of what represents the norm insofar as range of movement and end-feel is concerned.

What are the physiological and anatomical barriers which a particular joint should enjoy?

With that information, and a keen sense of end-feel (what the end of a movement should feel like compared with what is actually presented) should come an appreciation of what is needed in order to position a joint for receipt of MET input, irrespective of which joint is involved.

If end-feel is sharp or sudden, it probably represents protective spasm of joint pathology, such as arthritis.

The benefits of MET to such joints will be limited to what the pathology will allow; however, even in arthritic settings, a release of soft tissues commonly produces benefits. Kaltenborn (1985) summarises normal end-feel variations as follows:

- Normal soft end-feel results from soft tissue approximation (as in flexing the knee) or soft tissue stretching (as in ankle dorsiflexion).
- Normal firm end-feel is the result of capsular or ligamentous stretching (internal rotation of the femur for example).
- Normal hard end-feel occurs when bone meets bone as in elbow extension.

He defines abnormal end-feel variations as follows:

- A firm, elastic feel is noted when scar tissue restricts movement or when shortened connective tissue is present.

- An elastic, less soft end-feel occurs when increased muscle tonus prevents free movement.
- An empty end-feel is noted when the patient stops the movement, or requests that it be stopped, before a true end-feel is reached, usually as a result of extreme pain such as might occur in active inflammation, or a fracture, or because of psychogenic factors.
- As noted above, a sudden, hard end-feel is commonly due to interosseous changes such as arthritis.

By engaging the barrier (always the barrier, never short of the barrier for joint conditions) and using appropriate degrees of isometric effort, the barriers can be pushed back by means of our two physiological tools, postisometric relaxation and reciprocal inhibition, as the soft tissues are encouraged to release. Remember also Ruddy's 'pulsed MET' variation (p. 56) which is useful in joint problems.

Apart from those already discussed in this chapter, no specific joint guidelines are given because it is assumed that the reader can employ the principles as explained and the examples as given to adapt and extrapolate the use of MET methods to any/most joint conditions.

REFERENCES

Evjenth O, Hamberg J 1984 Muscle stretching in manual therapy. Alfta Rehab, Alfta, Sweden
Gelb H 1977 Clinical management of head, neck and TMJ pain and dysfunction. W B Saunders, Philadelphia
Goodridge J 1981 Muscle energy technique. Journal of the American Osteopathic Association 81: 249
Greenman P 1989 Principles of manual medicine. Williams and Wilkins, Baltimore
Grieve G 1986 Modern manual therapy. Churchill Livingstone, London
Harakal J 1975 An osteopathically integrated approach to whiplash complex. Journal of the American Osteopathic Association 74: 941–956
Hartman L 1985 Handbook of osteopathic technique. Hutchinson, London
Janda V 1988 In: Grant R (ed) Physical therapy of the cervical and thoracic spine. Churchill Livingstone, New York
Kaltenborn F 1985 Mobilisation of extremity joints. Olaf Norlis Boekhandel, Norway
Lewit K 1985 The muscular and articular factor in movement restriction. Manual Medicine 1: 83–85
Steiner C 1994 Osteopathic manipulative treatment – what does it really do? Journal of the American Osteopathic Association 94(1): 85–87
Stiles E 1984a Manipulation – a tool for your practice? Patient Care 18: 16–42
Stiles E 1984b Manipulation – a tool for your practice. Patient Care 45: 699–704
Yates S 1991 Muscle energy techniques. In: DiGiovanna E (ed) Principles of osteopathic manipulative techniques. Lippincott, Philadelphia

7

Integrated neuromuscular inhibition technique (INIT)

It is clear from the work of Travell and Simons in particular that myofascial trigger points are a primary cause of pain, dysfunction and distress of the sympathetic nervous system (Travell & Simons 1986).

Melzack and Wall (1988) in their pain research have shown that there are few chronic pain problems where myofascial trigger point activity is not a key feature maintaining or causing chronic pain.

As noted in Chapter 2, there are numerous causes for the production and maintenance of myofascial triggers, including postural imbalances (Barlow 1959, Goldthwaite 1949), congenital factors – warping of fascia via cranial distortions (Upledger 1983), short leg problems, small hemi-pelvis etc. – occupational or leisure overuse patterns (Rolf 1977), emotional states reflecting into the soft tissues (Latey 1986), referred/reflex involvement of the viscera producing facilitated segments paraspinally (Beal 1983, Korr 1976), as well as trauma.

LOCAL FACILITATION

According to Korr, a trigger point is a localised area of somatic dysfunction which behaves in a facilitated manner, i.e. it will amplify and be affected by any form of stress imposed on the individual whether this is physical, chemical or emotional (Korr 1976).

A trigger point is palpable as an indurated, localised, painful entity with a reference (target) area to which pain or other symptoms are referred (Chaitow 1991a).

Muscles housing trigger points can frequently be identified as being unable to achieve their normal resting length using standard muscle evaluation procedures as described in Chapter 4 (Janda 1983). The trigger point itself always lies in hypertonic tissue and not uncommonly in fibrotic tissue, which has evolved as the result of exposure of the tissues to diverse forms of stress.

Treatment methods

A wide variety of treatment methods have been advocated in treating trigger points, including inhibitory (ischaemic compression) pressure methods (Nimmo 1966, Lief 1982/1989), acupuncture and/or ultrasound (Kleyhans and Aarons 1974), chilling and stretching of the muscle in which the trigger lies (Travell and Simons 1986), procaine or xylocaine injections (Slocumb 1984), active or passive stretching (Lewit 1992), and even surgical excision (Dittrich 1954).

Clinical experience has shown that while all or any of these methods can successfully inhibit trigger point activity short-term, in order to completely eliminate the noxious activity of the structure, more is often needed.

Travell and Simons have shown that whatever initial treatment is offered to inhibit the neurological overactivity of the trigger point, the muscle in which it lies has to be made capable of reaching its normal resting length following such treatment or else the trigger point will rapidly reactivate.

In treating trigger points the method of chilling the offending muscle (housing the trigger) while holding it at stretch in order to achieve this end was advocated by Travell and Simons, while Lewit recommends muscle energy techniques in which a physiologically induced postisometric relaxation (or reciprocal inhibition) response is created, prior to passive stretching. Both methods are commonly successful, although a sufficient degree of failure occurs (trigger rapidly reactivates or fails to completely 'switch off') to require investigation of more successful approaches.

One reason for failure may relate to the possibility of the tissues which are being stretched not being the precise ones housing the trigger point.

Hypothesis

The author hypothesises that partial contraction (using no more than 20 to 30% of patient strength, as is the norm in MET procedures) may sometimes fail to achieve activation of the fibres housing the trigger point being treated, since light contractions of this sort fail to recruit more than a small percentage of the muscle's potential.

Subsequent stretching of the muscle may therefore only marginally involve the critical tissues surrounding and enveloping the myofascial trigger point.

Failure to actively stretch the muscle fibres in which the trigger is housed may account for the not infrequent recurrence of trigger point activity in the same site following treatment. Repetition of the same stress factors which produced it in the first place could undoubtedly also be a factor in such recurrence – which emphasises the need for re-education in rehabilitation. A method which achieved precise targeting of these tissues (in terms of tonus release and subsequent stretching) would be advantageous.

Selye's concepts

Selye has described the progression of changes in tissue which is being locally stressed. There is an initial alarm (acute inflammatory) stage, followed by a stage of adaptation or resistance when stress factors are continuous or repetitive, at which time muscular tissue becomes progressively fibrotic, and as we have seen in earlier chapters (Chapter 2 in particular), if this change is taking place in muscle which has a predominantly postural rather than a phasic function, the entire muscle structure will shorten (Janda 1985, Selye 1984).

Such hypertonic and perhaps fibrotic tissue, lying in altered (shortened) muscle, may not be easily able to 'release' itself in order to allow the muscle to achieve its normal resting length (as we have seen, this is a prerequisite of normalisation of trigger point activity). Along with various forms of stretch (passive, active, MET, PNF etc.), it has been noted above that inhibitory pressure is commonly employed in treatment of trigger points.

Such pressure technique methods (analogous to acupressure or shiatsu methodology) are often successful in achieving at least short-term reduction in trigger point activity and are variously dubbed 'neuromuscular techniques' (Chaitow 1991b).

Application of inhibitory pressure may involve elbow, thumb, finger or mechanical pressure (a wooden rubber tipped T-bar is commonly employed in the US) or cross-fibre friction. Such methods are described in detail in a further text in this series (*Modern Neuro-Muscular Techniques*).

TARGETING USING INTEGRATED NEUROMUSCULAR INHIBITION TECHNIQUE (INIT)

By combining the methods of direct inhibition (pressure mildly applied, continuously or in a make and break pattern) along with the concept of strain/counterstrain and MET, a specific targeting of dysfunctional soft tissues can be achieved (Chaitow 1994).

Strain/Counterstrain (SCS) briefly explained

Jones (1981) has shown that particular painful 'points' relating to joint or muscular strain, chronic or acute, can be used as 'monitors' – pressure being applied to them as the body or body part is carefully positioned in such a way as to remove or reduce the pain felt in the palpated point.[1]

When the position of ease is attained (using what is known as 'fine tuning' in SCS jargon) in which pain vanishes from the palpated monitoring tender point, the stressed tissues are felt to be at their most relaxed – and clinical experience indicates that this is so, since they palpate as 'easy' rather than having a sense of being 'bound' or tense (see Ch. 3, pp 47 & 48, for more detailed discussion of this phenomenon).

SCS is thought to achieve its benefits by means of an automatic resetting of muscle spindles – which help to dictate the length and tone in the tissues. This resetting apparently occurs only when the muscle housing the spindle is at ease and usually results in a reduction in excessive tone and release of spasm. When positioning the body (part) in strain/counterstrain methodology, a sense of 'ease' is noted as the tissues reach the position in which pain vanishes from the palpated point.

INIT method 1

It is reasonable to assume, and palpation confirms, that when a trigger point is being palpated by direct finger or thumb pressure, and when the very tissues in which the trigger point lies are positioned in such a way as to take away the pain (entirely or at least to a great extent), that the most (dis)stressed fibres in which the trigger point is housed are in a position of relative ease.

The trigger point would then be receiving direct inhibitory pressure (mild or perhaps intermittent) and (using SCS methods) would have been positioned so that the tissues housing it are relaxed (relatively or completely).

Following a period of 20 to 30 seconds in this 'position of ease' – accompanied by inhibitory pressure – the patient would be asked to introduce an isometric contraction into the tissues housing the trigger (and currently 'at ease') and to hold this for 7 to 10 seconds – so involving the very fibres which had been repositioned to obtain the strain/counterstrain release.

Following the isometric contraction there would be a reduction in tone in these tissues (postisometric relaxation). These could then be gently stretched as in any muscle energy procedure as described previously, with the strong likelihood that specifically involved fibres would be stretched (See Fig. 7.1).

---

[1] These tender points, as described by Jones, are found in tissues which were short rather than being stretched at the time of injury (acute or chronic) and are usually areas in which the patient was unaware of pain previous to their being palpated. They seem to equate in most particulars with 'ah shi' points in traditional Chinese medicine.

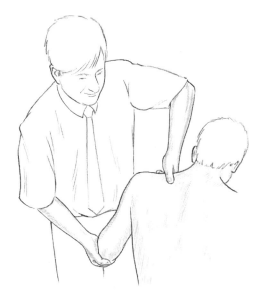

Figure 7.1A   First stage of INIT in which a tender/pain/trigger point in supraspinatus is located and ischaemically compressed, either intermittently or persistently.

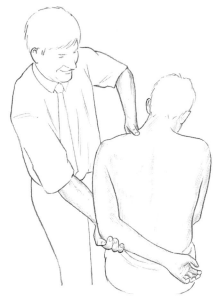

Figure 7.1C   Following the holding of the isometric contraction for an appropriate period, the muscle housing the point of local soft tissue dysfunction is stretched. This completes the INIT sequence.

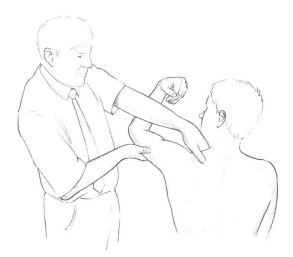

Figure 7.1B   The pain is removed from the tender/pain/trigger point by finding a position of ease which is held for at least 20 seconds, following which an isometric contraction is achieved involving the tissues which house the tender/pain/trigger point.

Method 2

There is another choice – a variation in which, instead of an isometric contraction followed by stretch being commenced following the period

of ease (strain/counterstrain position), an isolytic approach could be used.

The muscle receiving attention would be actively contracted by the patient at the same time that a stretch was being introduced – resulting in mild trauma to the muscle and the breakdown of fibrous adhesions between it and its interface and within its structures (Mitchell et al 1979).

To introduce this method into trigger point treatment, following the application of inhibitory pressure and SCS release, the patient would be asked to contract the muscles around the palpating thumb or finger (lying on the now inhibited pain point) with the request that the contraction should not be a full strength effort, since the operator intends to gently stretch the tissues while the contraction is taking place.

This isotonic eccentric effort – designed to reduce contractions and break down fibrotic tissue – should target precisely the tissues in which the trigger point being treated lies buried. Following the isolytic stretch the tissues could benefit from effleurage and/or hot and cold applications to ease local congestion. An instruction should be given to avoid active use of the area for a day or so.

Summary

The integrated use of inhibitory pressure, strain/counterstrain and one or other form of muscle energy technique, applied to a trigger point or other area of soft tissue dysfunction involving pain or restriction of range of motion (of soft tissue origin), is a logical approach since it has the advantage of allowing precise targeting of the culprit tissues.

Clearly, the use of an isolytic approach as part of this sequence will be more easily achieved in some regions rather than others – upper trapezius posing less of a problem in terms of positioning and application than might quadratus lumborum.

REFERENCES

Barlow W 1959 Anxiety and muscle tension pain. British Journal of Clinical Practice 13: 5

Beal M 1983 Journal of the American Osteopathic Association (July)

Chaitow 1991a Palpatory literacy. Harper Collins, London

Chaitow L 1991b Soft tissue manipulation

Chaitow L 1994 INIT in treatment of pain and trigger points. British Journal of Osteopathy XIII: 17–21

Dittrich R 1954 Somatic pain and autonomic concomitants. American Journal of Surgery

Goldthwaite J 1949 Essentials of body mechanics. Lippincott, Philadelphia

Janda V 1983 Muscle function testing. Butterworths, London

Janda V 1985 In: Glasgow E (ed) Aspects of manipulative therapy. Chuchill Livingstone

Jones L 1981 Strain/counterstrain. Academy of Applied Osteopathy, Colorado Springs

Kleyhans, Aarons 1974 Digest of Chiropractic Economics (September)

Korr I 1976 Spinal cord as organiser of the disease process. Yearbook of the Academy of Applied Osteopathy 1976

Latey P 1986 Muscular manifesto. Latey, London

Lewit K 1992 Manipulation in rehabilitation of the locomotor system Butterworths, London

Lief S 1982/9 Described by Chaitow in: Neuromuscular Technique 1982 which became Soft Tissue Manipulation 1989 (further revised in 1991). Thorsons, Wellingborough

Melzack R, Wall P 1988 The challenge of pain. Penguin, New York

Mitchell F, Moran P, Pruzzo N 1979 Evaluation of osteopathic muscle energy procedure. Valley Park, Missouri

Nimmo R 1966 Receptor tonus technique. Lecture Notes

Rolf I 1977 The integration of human structures. Harper and Row, New York

Selye H 1984 The stress of life. McGraw Hill, New York

Slocum J 1984 Neurological factors in chronic pelvic pain. American Journal of Obstetrics and Gynaecology 49: 536

Travell J, Simons D 1986 Trigger point manual. Williams and Wilkins, Baltimore

Upledger J 1983 Craniosacral therapy. Eastland Press, Seattle

8

Results of MET

The examples given in this short chapter relate to the application of MET in very different situations – chronic muscle pain, fibromyalgia, low back problems and acute joint problems following internal bleeding.

The sources are all derived from peer-reviewed literature relating to work carried out in settings as diverse as Polish, Swedish and Czech medical hospitals as well as a Southside Chicago osteopathic hospital.

The diverse spectrum of dysfunction involved mirrors much that is faced on a daily basis by therapists and practitioners, and it is hoped that review of these reports will encourage the wider use of the variety of MET possibilities as outlined in this and earlier chapters.

MET RESULTS IN TREATMENT OF MYOFASCIAL PAIN

David Simons, co-researcher with Janet Travell into trigger points, and Karel Lewit, the Czech developer of gentler MET, have fairly conclusively demonstrated the efficiency of MET in a study involving assessment and treatment of severe muscular pain, using MET.

244 patients with pain diagnosed as musculo-skeletal in nature were examined and found to have between them 351 muscle groups requiring attention, based on their having:

1. Trigger points in the muscle and/or its insertion
2. Increased muscular tension during stretch
3. Muscular-tension shortening which was not secondary to movement restriction caused by joint dysfunction.

These were muscular/soft tissue restrictions and not joint problems.

The method used in treating these muscles involved Lewit's postisometric relaxation approach as described in Chapters 3 (p. 50) and 4 (p. 66), in which prolonged but mild isometric contractions against resistance were carried out for 10 seconds before releasing. Following complete 'letting go' by the patient, and on a subsequent exhalation, any additional slack was taken up and the muscle moved to its new barrier ('stretch was stopped at the slightest resistance').

From the new position, the process was repeated, although if no release was apparent contractions, which remained mild, were extended for up to 30 seconds. It was often noted that it was only after the second or third contraction that a release was obtained, and 3 to 5 repetitions were usually able to provide as much progress as was likely at one session.

When release was noted the operator was careful not to move too quickly: 'At this time the operator was careful not to interfere with the process and waited until the muscle relaxed completely. When the muscle reached a full range of motion the tension and the tender (trigger) points in the muscle were gone.'

The results were impressive, with 330 (94%) of the 351 muscles or muscle groups treated, demonstrating immediate relief of pain and/or tenderness.

The technique was required to be precise concerning the direction of forces, which needed to be aligned to stretch the fibres demonstrating greatest tension. The patient's effort therefore needed to involve contraction in the direction which precisely affected these fibres. This was most important in triangular muscles such as pectoralis major and trapezius.

At a 3 month follow-up, lasting relief of pain was found to have been obtained in 63% (referring to the pain originally complained of) and lasting relief of tenderness (relating to relief of tenderness in the treated muscles) in a further 23% of muscles.

Among the muscles treated in this study those which were found to respond most successfully are given in Table 8.1.

The authors point out that, 'The technique not only abolished trigger points in muscles, but also relieved painful ligaments and periosteum in the region of attachment. The fact that increasing the length of shortened muscles relieved tenderness and pain, supports a muscular origin of the pain.'

The authors further point out that those patients achieving the greatest degree of long-term relief were those who carried out home treatment using MET stretches under instruction.

Table 8.1 Results of use of MET in myofascial pain study

Muscle	Number treated	Pain relief	Tenderness relief	No relief
Upper trapezius	7	7		
Wrist and finger flexors	5	5		
Lateral epicondyle of arm involving supinator, wrist and finger extensors and/or biceps brachii	20	19	1	
Suboccipital	23	21	2	
Soleus (Achilles tendon)	6	5	1	
Sternomastoid	9	7	2	
Hamstrings	8	5	2	1
Pelvic muscles/ligaments	29	22	4	3
Gluteus maximus (coccyx attachment)	27	15	9	3
Levator scapulae	19	10	7	2
Piriformis	21	11	6	4
Erector spinae	28	13	12	3
Deep paraspinal	15	7	5	3
Upper pectorals	22	10	5	7
Biceps femoris (fibula head)	18	8	6	4
Biceps femoris (long head)	7	2	0	5

MET RESULTS IN TREATMENT OF FIBROMYALGIA

Drs Stotz and Kappler of the Chicago College of Osteopathic Medicine have treated patients with fibromyalgia (Stotz & Kappler 1992) utilising a variety of osteopathic approaches including MET. The results given below were achieved by incorporating MET alongside positional release methods and a limited degree of more active manipulation (personal communication to the author, 1994).

Fibromyalgia is notoriously unresponsive to standard methods of treatment and continues to be treated, in the main, by resort to mild antidepressant medication, despite many of the primary researchers' insistence that depression is a result, rather than a cause of the condition in most instances (Block 1993, Duna & Wilke 1993).

The Chicago physicians measured the effects of osteopathic manipulative therapy (OMT – which includes MET as a major element) on the intensity of pain reported from tender points in 18 patients who met all the criteria for fibromyalgia syndrome (FMS) (Goldenberg 1993).

Each patient had 6 visits/treatments and it was found over a 1 year period that 12 of the patients responded well in that their tender points became less sensitive (14% reduction against a 34% increase in the 6 patients who did not respond well). Activities of daily living were significantly improved and general pain symptoms decreased.

In another study, 19 patients with all the criteria of FMS were treated once a week for 4 weeks at Kirksville, Missouri, College of Osteopathic Medicine using OMT which included MET as a major component. 84.2% showed improved sleep patterns, 94.7% reported less pain and most patients had fewer tender points on palpation (Rubin et al 1990).

MET RESULTS IN TREATMENT OF LOW BACK PAIN

Harald Brodin, of the Karolinska Hospital in Stockholm, describes the effects of using MET in a group of long-term, low back pain (lumbar area only) sufferers, specifically excluding patients with signs of disc compression, spondylitis or sacroiliac lesions, but not those with radiographic evidence of common degenerative signs – spondylosis deformans (Brodin 1987).

41 patients (24 female, 17 male) who had suffered pain in one or two lumbar segments, with reduced mobility, for a duration of at least 2 months, were randomly assigned to two groups, one receiving no treatment and the other receiving MET treatment, 3 times weekly for 3 weeks (see details below of the approach, which is described as 'a modification of the technique described by Lewit ... a variation of Mitchell's MET').

Both groups of patients recorded their pain level at rest and during activity according to a 9-graded scale each week.

After 3 weeks the group receiving treatment showed pain reduction statistically greater than in the non-treated group, as well as an increase in mobility of the lumbar spine.

Of the treated group, 4 remained the same or were worse, while 17 were improved, of whom 7 became totally pain free.

Only one in the non-treated group became totally pain free, while 16 remained the same or were worse. A total of 4 of this group, including the one who was totally improved, showed some improvement.

What was the treatment offered in this study?

This was divided into two phases, inhibition and facilitation, as follows:

1. Inhibition

- The patient was sidelying with the lumbar spine rotated by moving the upper shoulder backwards, with the tableside shoulder drawn forwards until the restricted segment of the spine was engaged.
- The operator stabilised the patient's pelvis as they pushed their shoulder forwards with a very small amount of effort against resistance from the operator for 7 seconds.
- During relaxation, the operator increased the degree of lumbar rotation to the new barrier, and repeated the isometric resistance phase again, until no further gain was made, usually 4 to 5 times in all.

2. Facilitation

- Active, rhythmic, small rotatory movements against the resistance barrier were carried out (see notes on Ruddy's approach in Ch. 4, p. 56).
- The patient held a deep breath and turned the head and looked in the direction of the rotation (towards the barrier).
- There was an isometric attempt, against the barrier to reduce this, to increase rotation.

Additional information given by the author of this study includes the fact that sometimes rotation away from the barrier was the starting point for treatment, with rotation towards the pain-free direction of motion. It is not noted when this decision (to move away from the barrier at the outset) was made but it is logical to assume that it was adopted when engagement of the restriction barrier was painful.

Patients were advised to use pain-free movements and positions during everyday life.

The author states, 'From this study we can conclude that in preselected cases, muscle energy technique is an effective treatment for lower back pain.'

Direct attention to soft-tissue imbalances might therefore be seen to be a means of helping to normalise painful spinal joints in many instances, especially if 'mobility is decreased, or its end-feel abnormally distinct' (Brodin 1982).

MET TREATMENT OF JOINTS DAMAGED BY HAEMOPHILIA

Just how useful MET can be in treating joint problems in even severely ill patients is illustrated by a Polish study of the effects of use of MET in a group of haemophiliac patients in which bleeding had occurred into the joints. There had also been bleeding into muscles such as iliopsoas, quadriceps and gastrocnemius (Kwolek 1989).

The study notes that, 'As a result of haemorrhage into the joints and muscles the typical signs and symptom of inflammation develop; if they are untreated or treated incorrectly or rehabilitation is neglected, motion restriction, deformation, athrodesis, muscle atrophy, scarring and muscular contractures may occur.'

Standard medical treatment used included electromagnetic field applications, heat, paraffin baths and massage as well – where appropriate – as the use of casts for limbs and other medical and surgical procedures.

All patients received instruction as to self-application of breathing, relaxation and general fitness exercise as well as rehabilitation methods for the affected joints using postisometric relaxation methods (MET). These were performed twice daily, for a total of 60 minutes. Range of movement was assessed and it was found that those patients using PIR (MET) methods achieved an improvement in range of movement of between 5° and 50° in 87% of the 49 joints treated – mainly involving ankles, knees and elbows (there was a reduction in motion range of 5° to 10° in just 6 joints). These undoubtedly impressive results for MET in a group of severely ill and vulnerable patients further highlights the safety of the method, since anything approaching aggressive intervention in treating such patients would be contraindicated.

The researchers, having pointed to frequent complications arising in the course of more traditional approaches, concluded, 'The 87% improvement in movement range of 5° to 50° and the lack of complications when rehabilitating articulations with haemophiliac arthropathy speaks in favour of routine application of the post isometric relaxation methods for patients with haemophilia.'

REFERENCES

Block S 1993 Fibromyalgia and the rheumatisms. Controversies in Rheumatology 119(1): 61–78

Brodin H 1982 Lumbar treatment using MET. Osteopathic Annals 10: 23–24

Brodin H 1987 Inhibition-facilitation technique for lumbar pain treatment. Manual Medicine 3: 24–26

Duna G, Wilke W 1993 Diagnosis, etiology and therapy of fibromyalgia. Comprehensive Therapy 19(2): 60–63

Goldenberg D 1993 Fibromyalgia, chronic fatigue syndrome and myofascial pain syndrome. Current Opinion in Rheumatology 5: 199–208

Kwolek A 1989 Rehabilitation treatment with post-isometric muscle relaxation for haemophilia patients. Journal of Manual Medicine 4: 55–57

Lewit K, Simons D 1984 Myofacial pain: relief by post-isometric relaxation. Archives of Physical Medical Rehabilitation 65: 452–456

Rubin B, et al 1990 Treatment options in fibromyalgia syndrome. Journal of American Osteopathic Association 90(9): 844–845

Stotz A, Kappler R 1992 Effects of osteopathic manipulative treatment on tender points associated with fibromyalgia. Journal of the American Osteopathic Association 92(9): 1183–1184

Index